OCEAN STATION

Also by Michael R. Adams

Shipboard Bridge Resource Management

OCEAN STATION

Operations of the U.S. Coast Guard, 1940–1977

Michael R. Adams

NOR'EASTER PRESS
Eastport, Maine

ISBN: 978-0-9779200-1-3

Published by Nor'easter Press
6 Water Street
Eastport, Maine 04631
www.noreasterpress.com

All photographs are courtesy of the United States Coast Guard and the United States Coast Guard Historian's Office.

All photographs are unretouched.

Charts in this volume were prepared by
International Mapping
Ellicott City, Maryland

Line art in this volume was prepared by
Trade Design Group
Long Beach, New York

Printed on acid-free paper.

First printing January 2010

10 9 8 7 6 5 4 3 2 1

Manufactured in the United States of America

PREFACE

IN TIMES OF PEACE, the U.S. Coast Guard patrols the mighty coasts and inland waters of the United States and fulfills myriad duties onshore and offshore, at home and abroad. In times of war, the Coast Guard is additionally called on to stand with its fellow Armed Services.

Across four decades of the twentieth century, the Coast Guard continuously sent cutters and crews to positions hundreds of miles offshore to make and report weather observations, to provide navigation and communication services, and to be available for search-and-rescue work. These stations were established in the early 1940s as *weather observation stations* and *plane guard stations* and were later known as *ocean stations.*

The ocean stations were made obsolete by the development of electronic and inertial navigation systems and weather satellites and buoys and by the reliability and versatility of modern aircraft. Most of the cutters that patrolled the stations have been decommissioned, and nearly all of the Coast Guardsmen who served on the stations have left the sea.

But their story remains.

* * *

MANY OF THE CUTTERS mentioned in this text patrolled through three wars and decades on stations. Consequently, they were subjected to changes in weapons systems, electronics, meteorologic equipment, and small boats, as well as superstructure modifications. When, where, and why that happened is described in books that focus on Coast Guard cutters and is not treated herein.

Because of the changes to the cutters, their displacement tonnages changed as well. Nevertheless, the cutters are described in the text by length and tonnage to give some indication of their sizes and to permit comparison among them. However, a tonnage should be considered an average for a respective class and not specifically accurate for any one cutter at any one time.

All tonnages given for Coast Guard cutters are full-load displacement tons, whereas those given for merchant ships are gross tons unless otherwise noted.

In this text, the unit *mile* represents the unit *nautical mile.*

Unless otherwise noted, all times given are ship local time, an important point because many operations were complicated by darkness. There is nothing darker than an overcast night at sea.

Finally, the Bibliography provides all sources used in the research of this volume and is divided in two parts: a section for general sources and a section for sources specific to individual chapters. Endnotes are set at the ends of chapters.

ACKNOWLEDGMENTS

I WOULD LIKE to thank the staff of the National Archives, Washington, D.C., and the staff of the U.S. Coast Guard Historian's Office, Washington, D.C., for their great assistance during the course of research for this volume.

CONTENTS

1

OS

THE WIND, sudden and fierce, blew out the sails of the square-rigged ship. The seas steepened, and *Pamir* swung into the trough of the seas then foundered as her cargo of grain shifted to port.

Pamir was carrying cargo in the year 1957, yet the four-masted barque was not just laden: Her professional crew of thirty-five was accompanied by fifty-one cadets who were aboard to be trained for the merchant marine.

Pamir was *in extremis* in the middle of the North Atlantic Ocean, some 1,600 miles east by north of Bermuda, an equal distance northwest by north of the Cape Verde Islands, and 600 miles west by south of the Azores. Steaming to her aid was the United States Coast Guard Cutter *Absecon*.

Absecon was there, more than a thousand miles from the nearest U.S. port, because her crew and the crews of several other cutters were conducting routine patrols of ocean stations far from the coasts of America.

* * *

ON JANUARY 25, 1940, President Franklin D. Roosevelt approved a recommendation to establish the Atlantic Weather Observation Service. The two stations of this service were manned by the United States Coast Guard to observe and report ocean weather conditions and to keep a radio guard over commercial aircraft crossing the ocean. After the United States entered World War II, the Coast Guard was directed to establish two new stations to observe and report ocean weather conditions and to provide assistance to military aircraft in the forms of navigational aid, air–ground–air communication services, and rescue in the event of ditching. The first two stations were referred to as *weather observation stations*, whereas the next two stations were referred to as *plane guard stations*.[1]

Each of the early weather observation stations in the Atlantic was identified by number, whereas each of the early plane guard stations in the Atlantic was identified by letter. The station identification schemes and station ship duties would later be made uniform: On April 1, 1944, all U.S. plane guard stations in the Atlantic were redesignated by number, and after V-E Day (May 8, 1945), all U.S. station ships in the Atlantic assumed the duties of plane guard ships.[2]

A total of forty-four Allied stations existed during World War II—twenty-two in the Atlantic Ocean and twenty-two in the Pacific Ocean.[3] The wartime stations of the Atlantic were primarily manned by the Coast Guard, and the wartime stations of the Pacific were primarily manned by the United States Navy. In 1944, Navy-commissioned, Coast Guard–manned frigates began to operate on the Atlantic and Pacific stations.[4]

The ocean weather information collected by the crews of the station ships was used for various purposes including support of Allied military operations, such as NEPTUNE—the cross-channel assault phase of OVERLORD.[5] After the conclusion of the war, the total number of stations was greatly reduced, and the Coast Guard continued to patrol the extant stations in the Atlantic and the Pacific.

By virtue of international agreement in September 1946, the stations were termed *ocean stations* (simultaneously *ocean weather stations*), and all stations were identified by letter. The station names were phonetically referred to as Able, Baker, Charlie, Dog, and so forth, yet after introduction of a revised phonetic alphabet across the period 1951–1956, the station names were phonetically referred to as Alfa, Bravo, Charlie, Delta, and so on.

The Coast Guard manned the at-sea stations from 1940 through 1977. For most of this period (i.e., from 1951 through 1971), the service primarily manned six stations: four in the North Atlantic and two in the North Pacific. The North Atlantic stations were Bravo (formerly Baker; between Newfoundland and Greenland at 56°30' N, 51°00' W), Charlie (about 600 miles southeast of Bravo at 52°45' N, 35°30' W), Delta (formerly Dog; almost due east of Nova Scotia at 44°00' N, 41°00' W), and Echo (formerly Easy; between Bermuda and the Azores at 35°00' N, 48°00' W). The North Pacific stations were November (formerly Nan; between San Francisco and Honolulu at 30°00' N, 140°00' W) and Victor (between Hawaii and Japan, first at 33°00' N, 164°00' E, then at 34°00' N, 164°00' E).

An ocean station was demarcated by an imaginary square drawn on the surface of the ocean that was 210 miles on each side (see diagram). This square was subdivided into a grid consisting of squares 10 miles on each side. The grid had a height of 21 squares and a width of 21 squares; thus, the grid was composed of 441 squares (i.e., 44,100 square miles). The grid was labeled alphabetically from A to V (north to south) along its left and right sides (latitude axes) and was labeled alphabetically from A to V (west to east) along its top and bottom (longitude axes). Thus, each square of the grid had a two-letter designation to denote its position within the grid, which position would be read beginning with latitude then longitude: The northwest-most square was AA, the next one east was AB, and so on. There were exceptions to the alphabetical labeling system of the grid: On the latitude axes, the letter S was deleted, and the letter O was transposed and inserted between letters J and K; on the longitude axes, the letter O was deleted, and the letter S was transposed and inserted between letters J and K. Thus, the square in the center of the grid—the central point of which represented the locus of the station coordinates—was designated OS: "On Station."

THE GRID. Shown is a reproduction of the 210-mile by 210-mile position-reporting grid issued by Capt. Miles H. Imlay, USCG (Air Sea Rescue Officer assigned to Commander Eastern Sea Frontier), on September 18, 1945. The grid was issued for use on U.S. Stations 1 to 11 in the North Atlantic. The grid was subsequently used on U.S. ocean stations in the North Atlantic. According to *Standard Operating Procedure for Pacific Ocean Stations* (Stika, 1950), the 210-mile by 210-mile grid was in use on U.S. ocean stations in the North Pacific by 1950; it was likely instituted before 1947.

The hatched square represents position OS ("On Station"). The central point of position OS is marked by the symbol "+"; this point represents the locus of the station coordinates (i.e., the station position). A cutter on an ocean station was obligated to keep station within the 100-square-mile space of position OS as close to the center of OS as possible. By the early 1960s, cutters on ocean stations were no longer obligated to keep station within position OS and were instead permitted to steam or drift within the central nine squares of the grid (i.e., within 15 to 21 miles of the center of OS). Prior to establishment of the grid, Coast Guard–manned vessels on weather observation stations or plane guard stations received various instructions concerning station-keeping; for instance, keep as close to the station position as possible; keep within a certain number of square miles of the station position; keep within a certain radius of the station position.

The symbol "×" appearing in position IM is not part of the grid but rather part of a navigational exercise that accompanied the newly issued grid, as follows: The published coordinates of the station are 56°30' N, 51°00' W. What is the bearing and distance of the cutter at position IM in relation to position OS? (North on the grid is true north. *Answer to Imlay's navigational exercise appears as a footnote at the end of Chapter 1.)

After 1956, position OS would be known by the phonetic Oscar Sierra. Source for grid: Imlay, 1945 (National Archives). Grid redrawn and reduced from original.

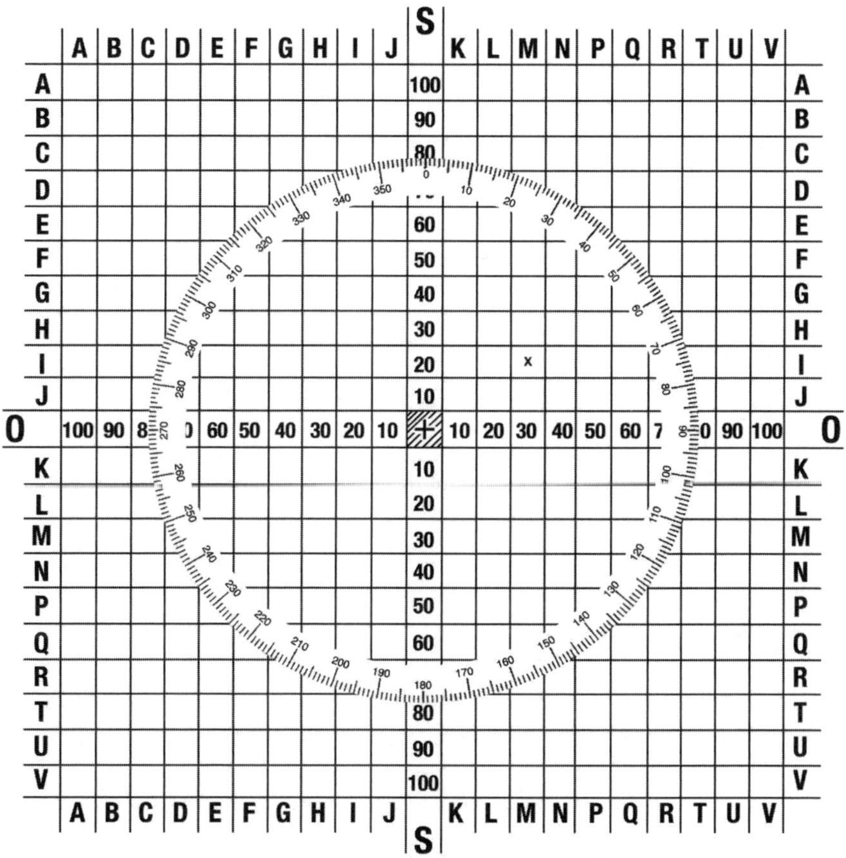

IT WAS COMMON PRACTICE for ocean station cutters to conduct relief of station at the edge of position OS.[6] Once the oncoming cutter arrived at the edge of OS, the pace aboard the cutter seemed to slow down with the engines. Whereas the cutter had sped along to effect relief, there was now no need to proceed on a full bell, as the cutter would generally be going nowhere for at least three weeks—the typical period of a patrol. Consequently, the cutter would steam along at slow speed while heading for the center of OS[7] as the relieved cutter disappeared over the horizon. Watching that other cutter depart could be poignant: It would often be the last close contact a crew had with other human beings until relieved of station.

The ocean station crews were essentially isolated. There were no satellite links; there were no cell phones; and there was no Internet, hence no e-mail. In fact, there was usually no mail at all. A cutter on station could try to make arrangements for the relief cutter to bring the mail that had stacked up at home, but this was logistically difficult because the cutters had homeports ranging from Portland, Maine, to Miami or from Long Beach, California, to Honolulu. Even if the relief cutter did bring mail, it was at least a week old when it arrived. Worse, if the weather was too nasty to complete a midocean transfer, then the mail would stay on the relief cutter until *it* got home after a three-week patrol.

Mail—when it could be delivered to station—was one of the few forms of contact the ocean station crews had with the rest of the world. Though radioteletype would enable the cutter's radio operators to copy international news broadcasts, those amounted to little more than headline summaries. Yet weather could isolate the crews from even this modest contact with the world; if atmospheric conditions prevented receipt of the radioteletype broadcasts, there would be no news at all.

A crewman on an ocean station was incommunicado with his family and friends, and, of course, they were incommunicado with him. The crewman was incommunicado with almost everyone except his shipmates. Some of the stations were so far off the shipping lanes (because they were along the air routes) that a cutter crew could complete an entire patrol without ever seeing another ship of any type until the relief cutter arrived. This had the effect of building a certain intense society within a cutter. The men were eating and sleeping and working next to one another for days and weeks on end with little other human contact.[8] Idiosyncrasies, skills (or lack thereof), interests, and aspirations were never secret.

The one place on a cutter that did have regular contact with the outside world was the combat information center (CIC). In fact, the CIC became the most active place aboard the cutter. The communications equipment and, after World War II, air-search radar displays that were located in the CIC were put into full-time use as the crewmen began their duties of tracking and providing navigational and weather information to overflying aircraft.

FLYING IN THE ERA of radio navigation, the pilot of each commercial plane (and the pilots of most military aircraft) approaching an ocean station would call in to the cutter and provide flight designation, point of departure, intended track, and estimated time of arrival at the destination. The men working in the CIC would be tantalized by reports of destinations such as London, Honolulu, Lisbon, and Tokyo while they slogged through days of watching the same part of the ocean pass by their cutter again and again.

If needed, the navigator of an aircraft could get a relatively accurate fix on the aircraft's position via the cutter and its radiobeacon. A cutter on an ocean station would broadcast station identification and cutter position four times hourly, three minutes each broadcast, in Morse code. Because at least eight numerals are required to show latitude and longitude in degrees and minutes, a cutter would instead broadcast a two-letter identification group assigned to the specific ocean station followed by a two-letter position group (i.e., the two-letter designation of the 10-mile by 10-mile square of the grid in which the cutter lay), with these groups followed by broadcast of a fifteen-second dash. Radiobeacon broadcast of position group OS represented to the aircraft navigator a position equating with the published coordinates of the identified station. Yet if the cutter was "in the grid" but at a position other than OS, the respective "off-station" position group would be broadcast, and the aircraft navigator would use the off-station position to plot the range and bearing of the cutter with respect to position OS (in plotting, a position was taken to be the exact center of the corresponding 10-mile by 10-mile square). By homing the aircraft on the radiobeacon signal, the aircrew could overfly the cutter and geographically mark the fix. Fixes by use of the grid system had a maximum error of 7.5 miles (likely probable error: 2.5 miles).[9]

Alternately, the aircrew could request an accurate range and bearing of the aircraft in relation to the cutter (obtained via the cutter's air-search radar) and the cutter's actual position in latitude and longitude (rather than just receiving the radiobeacon broadcast of the station identification and position groups) to obtain a high-quality fix. Also, the ground speed of the aircraft could be determined from the radar and that information passed along to the aircraft's navigator. Ground speed information was valuable over the ocean where an aircrew had no landmarks to use and when loran was unavailable for use.

Occasionally, an aircrew would report trouble that presaged ditching of the aircraft. Ditching is, simply, landing an aircraft on the water with no intention of returning it to the air. Though uncommon, ditching of aircraft on ocean stations did occur, particularly during the decade after World War II. The Coast Guard rescued 122 people from aircraft ditched on ocean stations, and the details of those rescues are the stuff of legend.

The CIC was the hub of ocean station duties not only because of the aviation-related activity: Weather observation balloons and radiosondes were launched by U.S. Weather Bureau personnel who were aboard or by crewmen, and the balloons were tracked on the radar in the CIC to obtain information on winds aloft. This data along with data on temperature, pressure, and humidity would be reported to shore-based forecasting stations or to the passing aircraft.

In the early 1960s, the ocean station crews began to conduct oceanographic probes while on station. A long wire would be lowered over the side of a cutter. Attached to the wire at specified intervals were metal cylinders with specialized thermometers, or Nansen bottles. When this Nansen cast was set, a weight, or *messenger*, was

Weather balloon launch.

released, slid down the wire, and tripped the first bottle, causing it to flip over, capture a seawater sample, record temperature, and release a new messenger, which slid down the wire to the second bottle, and so on. The bottles were retrieved, and temperature data and salinity data were transmitted via radioteletype to the U.S. Coast Guard Oceanographic Unit, which processed the data for the Navy's Fleet Numerical Weather Facility and for the National Oceanographic Data Center and later received the seawater samples and published related reports.

Thus, the watchstanders in the CIC and the bridge were kept relatively busy. The four-times-per-day weather balloon launches and once-per-day Nansen cast gave other crewmembers something different to do. But for most of the men aboard the cutter (the average crew size was around one hundred), often there was little of interest to do. When weather would permit, a cutter simply drifted in the center of station to hold position and conserve fuel. When weather did not permit, it was because a hurricane, typhoon, or other storm was battering the cutter, and the only thing most crewmembers could then do was hold on until the weather abated.

And the weather always took its toll. Storm after storm, rain, snow, freezing spray, howling winds, and heaping seas all beat the cutters and their crews. Equipment was destroyed and men were injured

Nansen cast: Working with a Nansen bottle.

as the cutters rolled, bucked, and plunged through one gale after another. Respites were few, particularly on the northernmost stations, and the calm days would find a crew cleaning up after a storm and restoring the gear that had been damaged. It seemed like a never-ending battle in a war against infinite forces.

The monotony might have been broken by routine maintenance, but painting is useless in cold climates and impossible in rain. Tearing apart machinery to conduct routine inspections was risky because no one knew when the cutter might have to proceed at full speed for search-and-rescue work. There would be drills to maintain proficiency, but the fact is that boredom was the bane of the ocean station crewmen.

All these things made the sight of the relieving cutter a thing of beauty. That cutter represented contact with the rest of the world. That cutter meant new newspapers, magazines, perhaps even mail. That cutter meant the Coast Guardsmen would be heading home.

That cutter was always white. The cutters of the U.S. Revenue Cutter Service, the Coast Guard's forerunner, had a long-standing tradition of being painted white. When the Revenue Cutter Service merged with the U.S. Life-Saving Service to become the U.S. Coast Guard in 1915, the tradition was carried on. Except for war years in the twentieth century when the cutters acquired coats of camouflage or gray paint, the offshore cutters always had and still have white hulls and superstructures.[10]

* * *

THE *CASCO*-CLASS CUTTER *Absecon* had been manning Ocean Station Echo in September 1957 as Hurricane Carrie formed off the Cape Verde Islands in the southern part of the North Atlantic. Carrie would have one of the longest hurricane tracks ever recorded, meandering toward Bermuda then veering to boil past the Azores almost to Ireland. On Thursday, September 19, the eye of Hurricane Carrie was approximately 300 miles west-southwest of the center of Ocean Station Echo, and the storm was headed *Absecon*'s way. Nevertheless, her crew remained on station. On Friday, September 20, the eye of Hurricane Carrie passed about 60 miles south of the center of Ocean Station Echo on a course toward the four-masted barque *Pamir.*

At 1526 GMT on Saturday, September 21, *Absecon* received from the ship *World Liberty* a relay of an SOS that had been broadcast earlier that day by *Pamir* ("SOS SOS SOS PAMIR").[11] At 1607 GMT, *Absecon* received from the ship *President Taylor* a relay of a distress message that had been broadcast by *Pamir* indicating the barque's position at 35°57' N, 40°20' W. At 1620 GMT, the cutter got under way. *Absecon*'s captain, Cdr. Ralph M. West, ordered up full speed, and the cutter departed OS on Ocean Station Echo bound for the foundering barque—a distance of about 350 miles. *Absecon*'s course was east by north with winds out of the north and seas from the northeast. This meant *Absecon* would pitch and roll into the seas throughout her transit, yet she kept making best speed.

Pamir was one of the largest sailing ships ever built at 3,020 tons and a hull length of 344 feet (375 feet overall). She had been built of steel in 1905 to round Cape Horn in the nitrate trade and had survived numerous encounters with heavy weather including Beaufort force 12 winds.[12] The Coast Guard cutter headed to help her was newer but smaller at 2,500 tons and 311 feet. *Absecon* was not the only ship that responded to *Pamir*'s signal for help. The ship closest to *Pamir,* however, was about six hours away.

PAMIR HAD SAILED from Buenos Aires on Saturday, August 10, 1957, bound for Hamburg with her professional crew, cadets, and cargo of grain. This mix of cargo and cadets came about through the Foundation *Pamir-Passat,* which was established in 1955 in then–West Germany to continue the tradition of training for sea by learning the ropes on a square-rigger. To defray the expenses of this effort, *Pamir*'s owners decided to load revenue-producing cargoes. Grain, a slow cargo, was a reasonable choice.

The loading of this cargo was the beginning of the end for the barque. *Pamir*'s owners had established a policy of loading of grain in bulk (i.e., loose) rather than stowing of bagged grain,[13] the latter method being the custom for years on large, grain-carrying sailing

Absecon in 1963, fitted out for ocean station patrols.

ships including *Pamir.* Because of a dockworkers' strike occurring in Buenos Aires, the barley was loaded by soldiers and trimmed by *Pamir*'s crew. The chief mate (usually in charge of cargo loading on commercial ships) had only two prior voyages on *Pamir,* his only time as an officer aboard a large sailing ship, and her master was making his first voyage in command of the barque.

Pamir's passage from Buenos Aires was uneventful for the first forty days. Then, late in the evening of Friday, September 20, *Pamir* began to encounter winds from the leading edge of Hurricane Carrie, which was approaching approximately west by south on a course that would put the barque in the northern perimeter of the hurricane and close to the eye. In the hours after dawn on Saturday, September 21, the powerful barque was sailing close-hauled on the starboard tack with twelve sails set[14] (six topsails, foresail, various staysails) in Beaufort force 9 winds[15] when caught by a sudden and dramatic increase in wind speed. The wind on the canvas forced *Pamir* into a 25-degree list to port. The order was given to shorten then to cut away sail. The wind continued to scream through the barque's rigging and began to blow out the sails and blew the furled sails from their gaskets. The seas became steep and sharp. Her sails lost, *Pamir* swung into the trough of the seas, and the 25-degree list to port continued because the cargo of grain had shifted to port. The list increased to 45 degrees, and the barque showed no intention of righting. The seas broke across *Pamir*'s decks, and water entered the hull because the barque was not properly battened before the storm.[16] *Pamir* lay beam to the seas and began to roll deeply.

Pamir's radioman managed to transmit the SOS and the distress message that were subsequently rebroadcast by other ships and received by *Absecon* and at least one other distress message. The second distress message received by *Absecon* was relayed by the ship *Tank Duke* at 1719 GMT as follows: "PAMIR AT 1401 GMT BT FOURMASTER BARQUE PAMIR/DKEF IN HEAVY HURRICANE POSN 35.57N 40.20W ALL SAILS LOST LISTING 45 DEGREES DANGER OF SINKING SIGNED MASTER."

Shortly before local noon at *Pamir*'s position (i.e., 1500 GMT), Captain Johann Diebitsch ordered all hands to put on life jackets and gave the order to abandon ship. *Pamir*'s list to port prevented lowering of the starboard lifeboats and subjected the port boats to pounding by the seas.

About twenty-one men were lowered in one lifeboat (boat no. 2); the bottom was holed during the launch, and the craft was swamped but kept afloat by its flotation tanks. A second lifeboat (boat no. 5) had its bow smashed during the attempted launch and was overturned by the seas, but ten men managed to right the craft in the water then get aboard; this lifeboat also was swamped and kept afloat by its flotation tanks.

The ship *President Taylor*, which was the ship closest to *Pamir*'s reported position and en route to assist, had established communications with the barque. *President Taylor*'s last contact with *Pamir* was at 1503 GMT on September 21, believed to be the time around which *Pamir* capsized.

After *Pamir* capsized, she proceeded to sink by the bow. Those in the two lifeboats saw a number of men in the water hanging onto a single life raft from *Pamir*. The rest of the cadets and crew were never seen again. Captain Diebitsch went down with his ship, as did all of *Pamir*'s officers and records.

At 2145 GMT, *President Taylor* was the first ship to arrive at the scene of *Pamir*'s last known position.

AT 0419 GMT ON SUNDAY, September 22, all ships within a 200-mile radius of *Pamir*'s last known position were advised to proceed to the distress scene to aid in the search for survivors.

The Canadian frigate HMCS *Crusader* arrived on scene at 0730 GMT and began to assign search areas to four merchant ships that were standing by. *Absecon* arrived in the search area at 1130 GMT and commenced a "creeping line search" in tandem with an aircraft. At 1522 GMT, *Absecon*'s lookouts sighted a raft. The cutter maneuvered alongside the raft only to find that it was one of two rafts that were reported at 0402 GMT to have been lost over the side by HMCS *Crusader* while the frigate was en route to *Pamir*'s last known position. The location of the raft, however, indicated that drift was in a south to southwest direction.

Ultimately, sixty merchant ships from thirteen nations and several aircraft from the U.S. Air Force and U.S. Navy as well as Portuguese aircraft would participate in the seven-day search for *Pamir*'s crew, with approximately ten to fifteen ships searching each day. Such a search must be coordinated or else ships and aircraft will be duplicating efforts and covering some areas multiple times while other areas remain unsearched, and search coordination is where the expertise of the Coast Guardsmen came into play. *Absecon* assumed control from HMCS *Crusader* at 1613 GMT on September 22 and became the on-scene commander. Arranging communications, assigning search areas, keeping track of who was where, and informing those ashore of progress were part of the duties *Absecon*'s crew carried out as they continued to look for *Pamir*'s crew.

THE NUMBER OF MEN in each of *Pamir*'s lifeboats decreased. Some were carried overboard by the seas, and, by the second day in the small vessels, some became overwhelmed and leaped overboard to their deaths.

During their ordeal, the men in the lifeboats sighted ships many times, but the crews of the ships failed to see them.

The lifeboat from *Pamir* that had started with ten individuals (boat no. 5) remained adrift for more than two days. *Absecon* was one of the ships that had passed within visual distance of this lifeboat, but her crew had failed to see it. Finally, at 2030 GMT on Monday, September 23, the ship *Saxon* located the lifeboat and rescued its five survivors, who were then transferred to USNS *Geiger*, which had a doctor aboard.

Pamir's swamped, varnished-wood lifeboats had the appearance of seaweed floating on the surface of the ocean.[17] Nevertheless, that the crews of merchant ships and a crew specializing in search and rescue failed to sight a lifeboat is not remarkable, as a search for people or boats in wind-whipped seas is a challenge. The height of the seas will hide a boat for a greater period of time than it is revealed. Also, the wind rips the seas into huge patches of foam and flying spray making it difficult to see anything. A boat may be on the surface but unseen because it is mistaken for foam or spray or is obscured by the huge quantity of such scud.

This foam and spray is also blown against the searching ship. Anyone standing outside as a lookout will be pelted by this horizontally driven water and by any rain, either of which stings and abrades exposed skin. Goggles can protect the eyes from this abuse but will also limit visibility as they fog up. There is no such thing as waterproof clothing in these conditions. The wind will force water up sleeves, down collars, and through loose seams in clothes. The apparently minor annoyance of having wet clothes becomes a true distraction after a couple of hours, thus a lookout becomes ever less effective as time goes on.

Anyone inside the ship will have to peer through portholes that are inundated with rain and/or driven spray. Nor will these individuals necessarily be dry. The same wind that finds the defects in clothing seams will seek out torn gaskets on doors and portholes and drive water through them. Rust holes or the cracks in windows and poor welds will let water inside. Nothing is dry, visibility is atrocious, and crewmembers are injured or become seasick as the ship endures a never-ending corkscrew motion that flings everyone through three dimensions.

ON TUESDAY, SEPTEMBER 24, the seas in the search area began to flatten. At 1752 GMT, the ship *Nordses* located the life raft from *Pamir*. The raft was empty and surrounded by sharks.

Less than an hour later, *Absecon*'s lookouts sighted a lifeboat, and the cutter sailed up to it. All Captain West had to do was get his 2,500-ton cutter alongside a few-hundred-pound lifeboat in the wind and the running seas without smashing into and destroying the small boat or steaming over it and crushing anyone left alive.

The best method to accomplish this task is to get the ship slightly upwind of the small boat and maneuver the ship so that it drifts broadside to the wind and down onto the boat. A stopped ship will almost always make greater leeway (i.e., distance downwind) than will a small boat because the ship has more surface area exposed to the wind than does the boat. This technique succeeds without crushing the boat for three reasons: (1) Even though the ship, because of its larger size and greater surface area, is drifting down onto the boat, the differential in drift speeds is minor and thus the closing rate is slow; (2) as long as the ship is making very little (or, ideally, no) headway when the two finally come together, there is little or no fore-and-aft, or pitching, momentum transferred from the ship to the boat; and (3) the ship will finally come to rest with its nearly flat midships section against the nearly flat midships section of the boat, which means the force of contact is spread out over the maximum surface area.

Furthermore, this technique guarantees that once the ship and boat are alongside each other, they will remain that way, and there will be little damage to the boat. The ship provides a lee for the boat so that the boat's motion is minimized, and the lee also ensures that the boat will no longer drift at all. It is simply pinned against the side of the ship.

With what must have been excellent timing, speed control, and a thorough knowledge of his cutter's drift rate, Captain West brought *Absecon* alongside the drifting boat at 1855 GMT. It was *Pamir*'s boat no. 5, from which the ship *Saxon* had rescued five survivors.

At 1935 GMT, *Absecon*'s lookouts sighted another lifeboat, and the cutter headed for it. In this lifeboat from *Pamir* that had held about twenty-one men (boat no. 2), there remained only one survivor, Gunter Hasselbach, who had by then been adrift in the boat for more than three days. Before *Absecon* could get alongside the lifeboat, however, Hasselbach jumped out of the boat and began swimming for the cutter. *Absecon* immediately executed her man-overboard procedures and launched one of her own boats to recover him.

Throughout the afternoon of Saturday, September 25, various search vessels reported finding debris and many empty life jackets with strings tied.

The search continued but for naught and was terminated at 0248 GMT on Sunday, September 29.

Pamir's Gunter Hasselbach is brought aboard *Absecon*.

A WEST GERMAN board of inquiry concluded that *Pamir* sank for a number of reasons: (a) The master was not familiar enough with the vessel, particularly in the conditions he encountered; (b) the chief mate had limited experience under sail; (c) the cargo shifted, a problem likely resulting from the nonstandard stowing practice; (d) *Pamir* had not been sufficiently battened down so that when she heeled, she opened up to the seas; (e) *Pamir*'s sails had not been gotten in early enough in anticipation of the storm; and finally (f) the board concluded that had *Pamir*'s ballast tanks been filled with water instead of the less-dense load of grain, she might have listed to only 20 degrees rather than the fateful 45 degrees.[18]

After *Pamir* had set sail on her fateful passage and while still en route, the four-masted barque *Passat* (the second sailing ship of the Foundation *Pamir-Passat*[19]) was similarly loaded with bulk grain in Buenos Aires. During *Passat*'s return passage, the cargo shifted to port in heavy weather and endangered the barque. Aware of the recent fate of *Pamir*, the master of *Passat* flooded the grain-filled ballast tanks,[20] and the barque stabilized. *Passat* was towed to Lisbon where the shifted cargo was restowed, and her subsequent arrival at Hamburg on December 8, 1957,[21] marked not only the ultimate transoceanic voyage of commercial cargo under square sail but also the end of the era of commercial cargo–carrying sailing school ships of the merchant marine.[22]

ENDNOTES

1. Coast Guard historical sources conflict regarding assertion of a third weather observation station established northeast of Newfoundland between the time of establishment of the first two weather observation stations and the time of establishment of the first two plane guard stations. This assertion may relate to relocated Station 2 (see Chapter 2).

2. Dinsmore provides an April 1, 1944, date based on station reorganization by Commander, Task Force 24 (Rear Adm. Edward H. "Iceberg" Smith, USCG), U.S. Atlantic Fleet. U.S. Coast Guard Headquarters Public Information Division (1949) provides a time frame of reorganization "by June 1944" when the plane guard reporting system was administratively incorporated into the weather observation reporting system. U.S. Coast Guard Headquarters Public Information Division (1949) provides the "after V-E Day" time frame for uniformity of station ship duties.

See Dinsmore, R.P., Captain, USCG (Ret.), "Ocean Weather Ships 1940–1980." U.S. Coast Guard Historian's Office, Washington, D.C. Available at: http://www.uscg.mil/history/webcutters/rpdinsmore_oceanstations.asp. Accessed 1 June 2009.

3. U.S. Coast Guard Headquarters Public Information Division (1949).

4. Willoughby (1989, p. 130).

The *Tampa*-class cutter USCGC *Haida*, a veteran of the Bering Sea Patrol, would assume one of the first two stations established in the Pacific—Able (49°00' N, 148°00' W), in later years known as Ocean Station Papa (50°00' N, 145°00' W)—beginning in June 1943 and would make patrols on Able through March 1946. From July 1944 to the end of World War II, *Haida* and the Coast Guard–manned *Tacoma*-class frigate USS *Pocatello* would rotate on patrol of this station. Other Coast Guard–manned *Tacoma*-class frigates would arrive on Pacific stations from August 1944 (USS *Grand Forks*) through 1945 and on Atlantic stations from October 1944 (USS *Abilene*) through 1945.

5. Willoughby (1989, p. 128). For a detailed description of NEPTUNE, see *United States Naval Administration in World War II. United States Naval Forces, Europe. Volume V. The Invasion of Normandy: Operation* NEPTUNE. London: Commander, U.S. Naval Forces in Europe, 1946. Mss. #147-E. Available at the Naval History & Heritage Command website: http://www.history.navy.mil/library/online/comnaveu/comnaveu_index.htm. Accessed 1 June 2009.

6. According to the logs of ocean station cutters, by the early 1960s, cutters conducted relief of station at a perimeter equating with the edge of the central nine squares of the grid rather than at the edge of OS, and by the mid-1960s, the cutters conducted relief of station at the edge of the grid, which gave the relieved cutter a quicker passage home.

7. A cutter on an ocean station was obligated to keep station within the 100-square-mile space of position OS as close to the center of OS as possible. According to the logs of ocean station cutters, by the early 1960s, cutters on ocean stations were no longer obligated to keep station within position OS and were instead permitted to drift within the central nine squares of the grid (i.e., within 15 to 21 miles of the center of OS). Prior to establishment of the grid, Coast Guard–manned vessels on weather observation stations or plane guard stations received various instructions concerning station-keeping; for instance, keep as close to the station position as possible; keep within a certain number of square miles of the station position; keep within a certain radius of the station position.

8. Throughout this book, Coast Guard crews are referred to as *men*, which is not meant in the generic sense of *mankind*. Rather, the service did not per-

manently assign women to any of its cutters until 1977—the same year the last of the ocean stations was closed.

9. Imlay (1945).

10. The service mark was added to U.S. Coast Guard cutters beginning in 1967, coincident with the transfer of the Coast Guard from the Department of the Treasury to the newly formed Department of Transportation.

11. Details regarding the time of the foundering of *Pamir*, the texts of distress messages, the events involving the lifeboats, and the search-and-rescue effort are taken from the report U.S. Coast Guard Cutter *Absecon* (1957b). Sources for events leading up to the foundering of *Pamir* include U.S. Coast Guard Cutter *Absecon* (1957b), Paterson (2007), Parrott (2003), Moore (1957), and West German Federal Minister of Trade (1958).

12. Parrott (2003, p. 42).

13. Parrott (2003, pp. 30–32, 50–53). During a 1951–1952 shipyard period, the four-masted barques *Pamir* and *Passat* were retrofit with ballast tanks, auxiliary engines, transverse watertight bulkheads, and guides for shifting boards in preparation for a West German training ship–cargo program, which initial effort failed but was then taken up by the Foundation *Pamir-Passat*.

14. Paterson (2007).

15. Parrott (2003, p. 36).

16. An archive of material discovered in Bremen, Germany, related to the loss of *Pamir* contained a letter from the previous master of *Pamir* complaining to the owners about the condition of the barque's deck: "The deck is so badly corroded . . . that it is no longer possible to make it watertight." The wooden deck planks laid over the steel deck were said to buckle when it rained. No repairs were made (Paterson, 2007).

17. Observation in U.S. Coast Guard Cutter *Absecon* (1957b).

18. West German Federal Minister of Trade (1958).

19. *Pamir* and *Passat* had originally been built to carry cargo for the Flying P Line, which was owned and operated by Germany's House of Laeisz. For details, see Villiers (1953); Parrott (2003); and others.

20. Parrott (2003, p. 50); Schäuffelen (2005, p. 112).

21. Schäuffelen (2005, p. 112).

22. Villiers (1971, pp. 320–327).

*Answer to Imlay's navigational exercise: The cutter bears 56°30' true, 37 miles from 56°30' N, 51°00' W.

2

STATIONS 1 AND 2

35°38' N, 53°21' W; 37°44' N, 41°13' W

THE *SECRETARY*-CLASS CUTTER *Duane* was patrolling Station 2 in June 1941. On the morning of June 13, *Duane* received a distress message from the British steamship *Tresillian* reporting that the ship was being shelled by a U-boat southeast of Cape Race, Newfoundland, beyond the Grand Banks, and needed help. *Duane* immediately got up full steam and headed off toward the ship's reported position not to stop the attack—the United States was not yet at war with Germany—but to rescue survivors.

However, *Duane*'s commanding officer, Cdr. Albert M. Martinson, could take no chances. He started preparing his cutter for the possibility the U-boat would not recognize it for what it was—a non-belligerent. Captain Martinson had ammunition for the 5-inch guns brought up from the magazines to the ready service lockers and made sure the depth charges were prepared for use.

On the morning of June 14, *Duane* arrived at the place where *Tresillian* should have been, but *Duane*'s crew found nothing. The British ship had succumbed to the attack by *U-77*[1] and slipped beneath the waves. Her attacker was long gone.

Duane's crew began a search for survivors, working the cutter eastward to allow for the likely drift of debris and boats. Three U.S. Navy flying boats had been dispatched to the area and were aiding in the search for *Tresillian*'s crew.

At 1025, one of the aircrews sighted two of the ship's boats about twenty miles farther east than *Duane* had ventured and directed the cutter to the scene. When they got the men from the two boats aboard, *Duane*'s crew learned that the entire crew of *Tresillian* had been saved.[2]

It was the auspicious beginning of a notable duty on station: Search and rescue.

* * *

ON SEPTEMBER 9, 1942, the crewmen of the cutter *Muskeget*, a former freighter built in 1923, completed their regular weather observations and transmitted their routine weather reports. The next day: silence. Thereafter, more silence. On September 13, the cutter *Monomoy*, a former freighter built in 1918, arrived on station to relieve *Muskeget*, but there was no *Muskeget* to relieve. After two days of steaming around the station and repeated attempts at radio contact, *Monomoy* reported that *Muskeget* could not be found. All ships and aircraft in the area were then directed to search for *Muskeget*, but the searchers found nothing.[3]

Or, rather, the searchers found nothing of *Muskeget*. *Monomoy* would later report that during the patrol from September 13 to October 1, the crew detected at least twenty submarines within striking distance of the station every day. *Muskeget* and *Monomoy* had been serving as weather ships on relocated Station 2 at position 53°00' N, 42°30' W, just north of the shipping corridor between North America and the United Kingdom. That is to say, their assigned station was located just north of three convoy transit routes[4] in the hunting territory of U-boats.

One year later, having heard nothing from any of *Muskeget*'s crew of 10 officers (including a U.S. Public Health Service doctor), 107 enlisted men, and 4 U.S. Weather Bureau personnel, the Coast Guard officially declared the cutter lost and the crew dead. Years later, Kriegsmarine records would show that *U-755* sank *Muskeget*.[5]

The freighters *Manasquan*, *Manhasset*, *Menemsha*, *Monomoy*, and *Muskeget* and the large yachts *Nourmahal* and *Sea Cloud* took over the patrols of Stations 1 and 2 in 1942. Though retrofit for war, the vessels were slow and not as heavily armed as the purpose-built cutters they replaced, yet while on station, the crew of *Monomoy* would drive off an attacking U-boat in July 1942, the crew of *Manhasset* would support convoy ONS-5 during a running battle against fifty-one U-boats in May 1943, and the crew of *Menemsha* would sight and attack a U-boat in August 1943.

The duty was difficult: Weeks of midocean isolation, dangerous weather, monotonous steaming, and the added burden of exposure by radio transmission, limited armament, and pending attack from the depths—a call unlike any that had come to the cuttermen before.

WHEN THEY BEGAN TO CROSS the open ocean, mariners were subject to the apparent whims of the weather. Although mariners would quickly learn of phenomena such as the trade winds and the Gulf Stream, the age of sail would begin to pass into the age of steam before global patterns of ocean winds and currents were discovered and charted.

In 1842, Lt. Matthew Fontaine Maury, USN, newly appointed as superintendent of the Navy's Depot of Charts and Instruments, began a study of old logbooks: "the handwritten daily record of nearly every voyage made by a ship of the United States Navy since its founding."[6] This study led Maury to discover seasonal patterns of ocean winds and currents.

Becoming the superintendent of the newly built U.S. Naval Observatory in 1844, Maury continued his study of the logbooks, and the result of this research was published by the Navy in 1847–1848 as eight sheets of Maury's new *Wind and Current Charts* for the North Atlantic Ocean. Charts for the other oceans followed,[7] and in 1850, Maury's *Notice to Mariners* (the third and subsequent editions of which were titled *Explanations and Sailing Directions to Accompany the Wind and Current Charts*) was issued by an independent publisher. The *Wind and Current Charts* instituted a revolution in sea routes and transit times, as the charts provided mariners with knowledge of seasonal ocean winds and currents and of optimal sailing routes based on those winds and currents.

Maury's work would continue and expand as he and his staff began to obtain meteorologic and hydrographic data from captains of merchant ships, who provided the data in the form of *abstract logs* in exchange for sets of Maury's charts. This wealth of ocean data was used in development of various series of *Wind and Current Charts* including the series Storm and Rain Charts—the first chart of this series was for the North Atlantic and published in 1853. The Storm and Rain Charts showed the number of days each month in which rain, fog, calm, lightning and thunder, or storm had occurred in specific regions of the ocean. The regions were represented on the charts as contiguous square sectors that were 5° of latitude by 5° of longitude in size.[8] The charts also included information regarding the directions from which storms had come. The Storm and Rain Charts, however, could not be used to forecast weather conditions at sea in real time.

The development in the mid-1890s of the wireless telegraph, or radiotelegraph, by Guglielmo Marconi made at-sea weather reporting in real time possible. In the first decade of the twentieth century, radiotelegraph enabled ships hundreds of miles apart to communicate with each other, and ships at sea could communicate ship to shore; hence, ships at sea could provide real-time weather information to shore-based forecasters, who in turn could process the information and provide offshore weather forecasts to ships at sea.[9] This exchange of information—which mirrored the exchange of abstract logs and charts between merchant ship captains and the U.S. Naval Observatory—was of benefit to everyone: Mariners learned of dangerous weather and could avoid or at least prepare for it; individuals ashore received coastal storm warnings; and shippers had greater assurance of the safe and timely arrivals of their cargoes.

The era when mariners began to receive offshore weather forecasts was followed by the era when aviators began to fly across open water. The airspace through which their propeller-driven aircraft flew was low enough that the surface weather was essentially the aviators' weather. Thus, the weather information gleaned from ships that led to offshore weather forecasts was as important to aircraft crossing the oceans as it was to ships.

As long as there was mutual benefit from the transmission of weather information and the receipt of weather forecasts, ships at sea continued to report. But when World War II broke out in Europe in the first week of September 1939 with its immediate U-boat campaign, weather reporting by ships in the Atlantic became a dangerous practice. It is possible to take a bearing on a radio transmission. Steer a course along that bearing, and one's ship, airplane, or submarine will wind up at the source of the transmission. In other words, a ship transmitting on radio in wartime is painting a bull's-eye around itself.

Therefore, radio silence was imposed on merchant ships of the belligerents, and other merchant ships crossing the Atlantic also stopped sending weather messages by radio.

The U.S. Weather Bureau was able to obtain increased at-sea weather reports from U.S.-flag ships, yet there remained "large areas in the Atlantic from which no reports are received."[10] Thus, the Weather Bureau turned to the president of the United States for help.

As did Juan T. Trippe, the president of Pan American Airways.

WHEN THEY LOOKED TO FLY over open water, aviators were hindered by lack of reliable navigation. It was common practice for pilots in the early part of the twentieth century to use ordinary road maps for navigation, but if there existed no visual references, as is the case over open water, then the only means of navigation were simple dead reckoning and celestial navigation.

Dead reckoning (i.e., using the known course and estimated speed of a craft to figure out how far and to where one had traveled) becomes less accurate over time: The longer an airplane (or a ship) proceeds without a good navigational fix, the longer it is exposed to the effects of wind (and sea) currents. These currents may be slowing the craft down, speeding it up, or offsetting it from its intended course.

Celestial navigation (i.e., use of a sextant and a chronometer to get a fix) can be used for over-water flights but cannot be relied on, particularly not for regularly scheduled over-water flights, because of the risk, hence threat, of overcast skies.

In May 1919, a Navy aircrew commanded by Lt. Cdr. Albert C. Read and accompanied by a Coast Guard pilot, Lt. Elmer F. Stone, flew a Navy-commissioned flying boat designated NC-4 on a multi-leg flight from Rockaway, New York, to Plymouth, England, thus accomplishing the first trans-Atlantic flight. The Navy had deployed five battleships to serve as weather stations for the flight and had strung thirty-five destroyers, or "station ships," at fifty-mile intervals along the Newfoundland–Azores–Lisbon legs of the planned air route to serve as visual navigational waypoints for the aircraft both day and night.[11] The operation was, in effect, a daring proof of concept.

The practical solution to the problem of over-water aircraft navigation would later come via radio. Pan American Airways (Pan Am) began operations in October 1927 with a contract airmail route between Key West, Florida, and Havana, Cuba. The airline was able to make regularly scheduled over-water flights on this route irrespective of overcast because of a radio direction-finding device developed for Pan Am by Radio Corporation of America (RCA) engineer Hugo C. Leuteritz, a Navy-trained engineer who would become the airline's chief communications engineer in January 1929.[12] The device was a lightweight radio transmitter that incorporated a loop antenna. During a flight, a Pan Am pilot would use the loop antenna to home on a signal that was broadcast by a radiobeacon at the destination.[13]

Leuteritz would later develop an entirely different, longer-range radio-navigation system for Pan Am: This system was based on broadcast of a signal by the aircraft and receipt of the signal by specialized antenna arrays at dedicated shore stations where the source of the signal would be plotted (i.e., Marconi–Adcock high-frequency radio direction-finding) and a heading thereby radioed to the aircraft pilot[14] to augment celestial navigation.

Thus, radio navigation would enable Pan Am to begin regular crossing of the oceans. In November 1935, Pan Am inaugurated regularly scheduled, multi-leg trans-Pacific service, and in May and June 1939, Pan Am inaugurated regularly scheduled trans-Atlantic service along the New York–Azores–Lisbon and the Newfoundland–Ireland air routes, respectively.

The sudden lack of at-sea weather reporting in September 1939 was of immediate concern to Juan T. Trippe: Without ocean weather information, Pan Am's clippers could not safely fly transoceanic routes.[15] Hence, Trippe approached the Coast Guard, the Navy, and the White House to lobby for weather information supplied by ships stationed along the North Atlantic air routes.[16]

THE MATTER BECAME pressing in late 1939. The Weather Bureau was calling for midocean weather stations. The secretary of Agriculture and the secretary of the Navy sought the stations as well. And the president of Pan Am had stated he would need to suspend the airline's trans-Atlantic service if weather reports could not be received. Although the Coast Guard was the logical choice to man the stations due to its previous ocean weather observation projects, the service did not have enough cutters in its Atlantic offshore fleet to do the work.

Consequently, President Roosevelt's naval aide, Navy Capt. D.J. Callaghan, was dispatched to ask Adm. Harold R. Stark, chief of naval operations, Rear Adm. Russell R. Waesche, Sr., commandant of the Coast Guard, and Rear Adm. Emory S. Land, chairman of the U.S. Maritime Commission, to solve the dilemma.

The three admirals recommended that the Coast Guard end the Grand Banks Patrol,[17] which would free up cutters to serve temporarily as weather ships until such time that the Maritime Commission received funds from Congress to rehabilitate four old ships for the purpose.

The recommendation of the admirals was spelled out in a memorandum prepared by Captain Callaghan and presented to the president for approval. Roosevelt signed the document on January 25, 1940, thus establishing the Atlantic Weather Observation Service.[18] On January 27, Coast Guard Headquarters discontinued the Grand Banks Patrol and on February 1 issued orders to the Boston, New York, and Norfolk Districts to begin operations on Station 1 (35°38' N, 53°21' W) and Station 2 (37°44' N, 41°13' W), which positions were along the great circle route between Bermuda and the Azores (see Chart 1). The orders instructed cutters on Stations 1 and 2 to make and report weather observations, to keep a radio guard over aircraft flying the route, and to keep station as close to the station coordinates as possible.[19]

The Coast Guard sent two of its newest, longest range, most seaworthy cutters, *Bibb* and *Duane*—both 2,350-ton, 327-foot *Secretary*-class cutters—to assume these stations. After these first two patrols were completed, a rotation began among the rest of the cutters in the class that were still on the East Coast: *Hamilton*, *Spencer*, *Ingham*, and *Campbell* (*Taney* was then in the Pacific). Later, the six East Coast cutters were joined on the patrols by the *Tampa*-class cutter *Mojave* (1,950 tons, 240 feet). All of these cutters had substantial modern

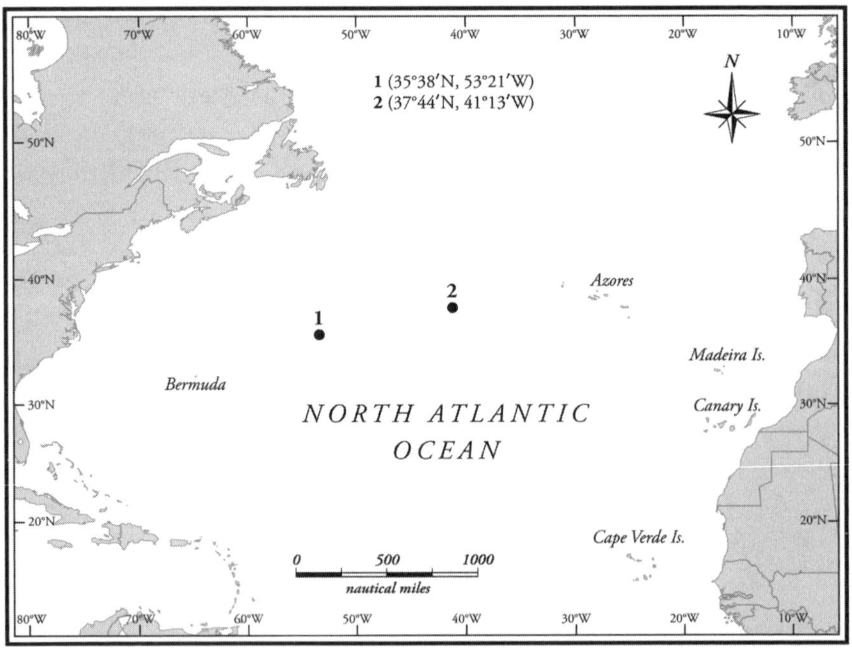

CHART 1. The chart shows the positions of Station 1 (35°38' N, 53°21' W) and Station 2 (37°44' N, 41°13' W) as originally established in February 1940. Sources: U.S. Coast Guard Cutter *Bibb*, Official Logbook, February 1940 (Station 1); U.S. Coast Guard Cutter *Duane*, Official Logbook, February 1940 (Station 2). See also Appendix I.

Chart prepared by International Mapping.

armament, which was advantageous when deploying them in what was essentially a war zone.

Twenty-one months after establishment of Stations 1 and 2, the Coast Guard was transferred from the Department of the Treasury to the Department of the Navy by Executive Order 8929 of November 1, 1941.[20]

* * *

SIX MONTHS AFTER the *Tresillian* incident, the United States entered World War II on December 8, 1941. In summer 1942, the Coast Guard was directed to establish two plane guard stations to serve U.S. bombers and fighters that would fly the North Atlantic Ferry Route via Greenland and Iceland as part of BOLERO (the buildup of U.S. forces in the United Kingdom). This ferrying of U.S. Army Air Forces combat aircraft out of reach of the U-boat wolf packs would support POINTBLANK (the combined American and British bomber offensive against the Third Reich).[21] One plane guard station, Station A, was established between Labrador and Greenland, and the other, Station B, was established between Greenland and Iceland.[22] The cutters on these stations were instructed to make and report weather observations, to serve as navigational waypoints for aircraft,[23] to provide air–ground–air communication services, and to rescue survivors should a plane be forced to ditch in the frigid North Atlantic waters. The cutters that patrolled Stations A and B did so as an extension of the Coast Guard's Greenland Patrol (commanded by then-Capt. Edward H. "Iceberg" Smith, USCG), which on October 25, 1941, had been reorganized as a task group under Commander, Support Force, Atlantic Fleet (Rear Adm. Arthur L. Bristol, Jr., USN). The Support Force was designated Task Force 4 on March 1, 1941, and ultimately as Task Force 24 on March 13, 1942.

The original Stations 1 and 2 would be repositioned. In January 1942, Station 1 was located at 39°00' N, 50°00' W, near both the Bermuda–Azores and New York–Azores flight paths. In January 1942, Station 2 was located at 52°00' N, 42°30' W, near both the Newfoundland–Scotland air route of the North Atlantic Ferry Route and the Newfoundland–Ireland air route. When *Muskeget* was torpedoed in September 1942, Station 2 was located at 53°00' N, 42°30' W.

In fall 1942, operational control of the cutters on weather observation stations was transferred from the Commandant of the Coast Guard to Commander, Task Force 24 (COMTASKFOR 24; Rear Adm. Roland M. Brainard, USN). On April 1, 1944, the plane guard stations were redesignated by number and incorporated with the weather observation stations into a task group comprising eight numbered stations—the Atlantic Weather Patrol—which remained under the operational control of COMTASKFOR 24 (Rear Adm. Edward H. "Iceberg" Smith, USCG).[24] The headquarters of COMTASKFOR 24 was located at Naval Operating Base Argentia.

BECAUSE THE COAST GUARD'S fleet of ten *Lake*-class cutters (2,075 tons, 250 feet) had been transferred to the Royal Navy in 1941 by way of the Lend-Lease Act, and because the Coast Guard was already using available cutters for short-range convoy escort, for antisubmarine patrols, and on the Greenland Patrol, and because by mid-1942, the *Secretary*-class cutters had been pulled from Stations 1 and 2 and were engaged in transoceanic convoy escort, and because *Mojave* had been reassigned to the Greenland Patrol, Coast Guard manning of the wartime stations would necessitate a hodgepodge fleet that included three classes of Coast Guard cutters (*Cactus*, *Active*, and *Algonquin* classes); trawlers; a gunboat (*Northland*); Navy-commissioned, Coast Guard–manned frigates (U.S. Navy *Tacoma* class); ex-merchant ships; and two large yachts.[25]

The *Cactus*-class cutters were comparatively small ships (1,000 tons, 180 feet) designed to set and recover buoys, which meant they had very low main decks. That these buoy tenders survived the North Atlantic's worst weather and successfully performed their mission is a tribute to the seaworthiness of their design and the seamanship of their crews.

These were not, however, the smallest cutters the Coast Guard sent out on the North Atlantic stations. That honor belonged to two of the *Active*-class cutters (*Active* and *Frederick Lee*), which had been built in the mid-1920s. These patrol boats displaced only 230 tons and were 125 feet in length, and each served on the Greenland Patrol.

That *Active* was not lost while patrolling Station A in the Davis Strait is notable when considering that much larger Coast Guard cutters would be heavily damaged by the weather in the strait in later years.

The *Algonquin*-class cutters were built between 1932 and 1934, and five of these were homeported on the East Coast: *Algonquin, Comanche, Escanaba, Mohawk,* and *Tahoma.* The *Algonquin* class had the same displacement as the *Cactus* class (1,000 tons) but only a 165-foot length. The heavy weight of the *Algonquin* class meant that these light icebreakers were slow, good for only thirteen knots, whereas the near-contemporary *Secretary*-class cutters could make twenty knots. Despite their small size, the *Algonquin*-class cutters proved themselves to be quite useful on the Greenland Patrol, on convoy escort, and on weather observation and plane guard station patrols in the North Atlantic.

The *Algonquin* class did not escape the war without loss. In June 1943, *Escanaba* and other cutters were escorting convoy GS-24 en route from Greenland to Newfoundland. At 0510 on June 13, *Escanaba* exploded violently from an undetermined cause. Crewmen aboard the cutter *Storis* (an icebreaking supply ship) saw huge yellow flames and a cloud of black smoke. Within three minutes, *Escanaba* was gone. *Storis* screened the *Arundel*-class cutter *Raritan* (an icebreaking tug) while the latter searched the scene, and the crew of *Raritan* found only two survivors and one body from *Escanaba*'s crew of 104 along with some debris.[26]

The cutter *Aivik* (590 tons, 128 feet) was one of ten steel-hulled and wood-hulled trawlers acquired on the waterfront by Captain Smith for service on the Greenland Patrol. The trawlers were retrofit and then commissioned as cutters from June through August 1942. Their Coast Guard crews, like the crews of other cutters on the Greenland Patrol, would go on to perform remarkable work both on and off the ice sheet, including resupply of the Bluie bases and stations. *Aivik* and other of the steel-hulled trawlers would patrol the plane guard stations,[27] as would the cruising gunboat *Northland* (2,150 tons, 216 feet)—a cutter commissioned in 1927 for Arctic operations that served as the flagship of Commander, Greenland Patrol.

The *Tacoma*-class frigates (2,200 tons, 303 feet; originally classified as gunboats) had been designed for the Navy for use in antisubmarine warfare. The keel of *Tacoma*, the first of the revised classification, was laid down in March 1943 at the Kaiser Cargo (Richmond No. 4) shipyard in Richmond, California, under a U.S. Maritime Commission contract, and the frigates began to enter service in early 1944. Because of the growing success of U.S. forces against the U-boat wolf packs in the Atlantic in the latter half of 1943, however, the *Tacoma*-class frigates were no longer urgently needed for Atlantic convoy and antisubmarine duty, thus the frigates and their Coast Guard crews were assigned to various tasks, which included convoy escort, amphibious operations, as well as service on the weather and plane guard stations of the Atlantic beginning in October 1944 (USS *Abilene*) and of the Pacific beginning in July 1944 (USS *Pocatello*). Ninety-six of the frigates were built during World War II; of these, seventy-five were manned by Coast Guard crews during the war, and twenty-one were transferred to the Royal Navy.[28] Sixteen of the frigates would be turned over by the Navy after the war and commissioned as Coast Guard cutters.

Two weather ships were former merchant ships that had each been converted into a 1940s version of a stealth antisubmarine warfare ship. USS *Big Horn*, a tanker, and USS *Asterion*, a freighter, looked like ordinary merchant ships but had substantial armament added. These so-called Q-ships, or "mystery ships," appeared to be valuable, unarmed targets to lurking U-boats, but their "cargo" consisted of heavy armament. The idea was simple: A Q-ship, appearing to be a lone ship or a convoy straggler, would draw an attack from a U-boat. The U-boats often surfaced and used their deck guns to destroy merchant ships to save their torpedoes for times when a submerged attack was required. Thus, if a U-boat surfaced and started to shoot at a Q-ship, it would be met not by an easy target but by an assault from guns and depth charges.

The Q-ships met with little success, however, and the ruse was over in November 1943. That month, *Big Horn* was in the midst of no less than nine U-boats, but none of them rose to the bait. *Big Horn*

The cutter *Big Horn*.

and *Asterion* were transferred to the Coast Guard, commissioned as cutters, and assumed duties as North Atlantic weather ships. *Big Horn* was later transferred back to the Navy to serve as an oiler in the Pacific.[29]

The most unusual vessel to sail as a weather ship was a square-rigger. A four-masted barque, *Sea Cloud* was launched at Kiel, Germany, in 1931 as a well-appointed private American yacht. As the war in Europe moved into its third year in late 1941, *Sea Cloud* was offered to the Navy, who decided to turn it over to the Coast Guard. At the Coast Guard Yard in Curtis Bay (Baltimore, Maryland), the barque was, to use the old sailor's jargon, cut down; all the yardarms were removed as were all but one of the masts. Replacing this weight were two 3-inch guns, four 20-mm guns, and some additional antisubmarine weapons. The alterations were complete in April 1942, and *Sea Cloud* was commissioned as a Coast Guard cutter and spent the next two years as a weather ship in the North Atlantic. By fall 1944, *Sea Cloud* was no longer needed for duty and reported to Boston where she was decommissioned and returned to her owners who restored the vessel to its former glory.[30]

Another of the weather ships, the cutter *Manasquan*, a *Lakes*-type freighter similar to *Menemsha* and *Monomoy*, would participate in testing of a developmental navigation system that would ultimately outlive the ocean stations.

IN OCTOBER 1940, physicist Dr. Alfred L. Loomis, chairman of the Microwave Committee of the then–National Defense Research Committee (NDRC), proposed to the Microwave Committee the development of a "pulsed, hyperbolic, radio grid-laying system" for the purpose of long-range navigation. The proposal was quickly accepted,

The cutter *Sea Cloud*.

and "Project 3" was assigned to a subcommittee, the Project 3 Committee, with project management provided by Bell Telephone Laboratories in New York.[31] On March 24, 1941, the secretary of the Treasury granted the NDRC permits to occupy two inactive U.S. Coast Guard lifeboat stations—one at Montauk, New York, and the other at Fenwick Island, Delaware—for purposes of experimentation.[32]

Project 3, however, was "foredoomed to failure under control of a loose administrative committee." By early summer 1941, the project was transferred from the subcommittee to the newly established Radiation Laboratory Navigation Group at the Massachusetts Institute of Technology, later joined by the Bell Telephone field staff who had worked on Project 3, and the original Project 3 was terminated.[33] The ensuing research and development work by the Radiation Laboratory Navigation Group showed potential of the technology for use in long-range navigation, which was formally brought to the attention of the U.S. Army and U.S. Navy in March 1942.

The Navy, in the person of Rear Adm. Julius A. Furer, coordinator of research and development for the secretary of the Navy, determined that the technology was worthy of further investigation. Observing that the Radiation Laboratory could benefit by some guidance and assistance to bring the technology to practical application and noting that any new aid to navigation developed would ultimately come under the authority of the Coast Guard, the Navy in May 1942 requested that the Coast Guard provide an officer with radio and electronics experience to be detailed for special duty to the chief of naval operations. Unknown to the Coast Guard, the officer would become the naval liaison officer at the Radiation Laboratory and in the field; known to the Coast Guard, the assignment "related to an electronic navigational device which might have war importance." That Coast Guard officer was Lt. Cdr. Lawrence M. Harding, and that developmental technology became known as loran.[34]

The system would consist of pairs of radio transmitting stations that would emit signals on a time-coordinated basis. The time differential (in microseconds) between the receipt of signals from a pair of transmitting stations (i.e., an initial "master" signal from one station and a later-transmitted, secondary "slave" signal from the other, paired station) measured with a loran receiver-indicator and then interpolated on a loran chart would give a navigator a hyperbolic *line of position*.[35] If a second pair of stations (or a second slave station) was introduced, then a second hyperbolic line of position was obtained that would intersect with the first to give a good navigational fix.

By June 1942, high-power transmitters had been installed at the Montauk and Fenwick Island stations, and Commander Harding arranged for June–July 1942 testing of the technology to answer two questions: (1) Could the signals from a single pair of stations be used to determine a single line of position? (2) What was the range of the signal? A Navy blimp stationed at Naval Air Station Lakehurst was outfitted with a new model of loran receiver-indicator, which was tested aloft on June 13, and the cutter *Manasquan*, outfitted with an identical receiver-indicator, was afterward sent on an extended at-sea navigational test. On July 17, 1942, Commander Harding received the results of the *Manasquan* test: The loran signal had an effective range offshore up to 680 miles in daytime and up to 1,300 miles at night and could be used to obtain a line of position.[36]

Coastal loran chains (i.e., series of three or four transmitting stations) then began to be built first in the Atlantic and then in the Pacific, and additional work began on preparation of loran charts, installation of onboard receiver-indicators, and increase of range of the signal. By enabling navigators to obtain accurate fixes across long ranges in all weather conditions, loran would help U.S. forces defeat the U-boat wolf packs in the Battle of the Atlantic by facilitating at-sea rendezvous of North Atlantic convoys, escort aircraft, and warships.[37]

FREIGHTERS AND FRIGATES, icebreakers and yachts, buoy tenders and patrol boats, trawlers and a tanker represent the variety of types of vessels the Coast Guard had to use to man the Atlantic stations during World War II.

Conversely, after World War II, the ocean station fleet primarily consisted of four classes of cutters: (1) the *Owasco*-class cutters (1,980 tons, 255 feet), which had been built for the Coast Guard in the period 1944–1946; (2) the *Secretary*-class cutters (2,350 tons, 327 feet); (3) the *Casco*-class cutters (2,500 tons, 311 feet), former Navy seaplane tenders acquired by the Coast Guard beginning in 1946; and finally (4) the *Hamilton*-class cutters (3,000 tons, 378 feet), designed and then built for the Coast Guard from 1965 to 1972. Sixteen ex-Navy *Tacoma*-class frigates and twelve ex-Navy *Edsall*-class destroyer escorts (1,490 tons, 306 feet) would be commissioned as Coast Guard cutters for discrete periods: the *Tacoma* class briefly after World War II and the *Edsall* class during the Korean War.

THE WORLD WAR II STATIONS were not only patrolled by multiple ship types but also manned by multiple nations. Brazil and the United Kingdom manned stations in the Atlantic, and Canada manned one station in the Pacific.[38]

Because of the usefulness of the stations, an international network of ocean stations would be established in the North Atlantic and a bilateral ocean station program would be established for the North Pacific as a world at peace began to resume normal commerce.

In March 1946, the Provisional International Civil Aviation Organization (PICAO; later the International Civil Aviation Organization, or ICAO) held a North Atlantic Regional Route Service Conference in Dublin, Ireland, to begin establishing rules and procedures for civil aviation across that ocean in the post-war world. In September 1946, the First Conference on North Atlantic Ocean Stations was held in London, England. On September 25, nine nations—Belgium, Canada, France, Ireland, Norway, Sweden, The Netherlands, the United Kingdom, and the United States—concluded the International Agreement on North Atlantic Ocean Weather Stations, which called for establishment of thirteen ocean stations in the North Atlantic to provide at-sea meteorologic, air navigation, communication, and search-and-rescue services.[39]

All stations would be identified by letter. Seven of the stations would be operated by the United States (Stations A, C, D, E, F, G, and H); the operation of one station would be shared by Canada and the United States (Station B); two stations would be operated by the United Kingdom (Stations I and J); the operation of one station would be shared by Belgium and The Netherlands (Station K); one station would be operated by France (Station L); and the operation of one station would be shared by Norway, Sweden, and the United Kingdom (Station M).[40] All the stations were to begin operation no later than July 1, 1947, and operate for three years. Finally, the agreement contained a provision for another conference to take place in 1949 so that the participants could evaluate the effectiveness and ongoing need for the stations.

Though the United States had assented to the agreement in principle, problems of implementation arose. Theoretically, there were enough ships to patrol the proposed stations because so many had been built for the Navy and the Coast Guard during the war; however, personnel and funding were in short supply.

After World War II, the Coast Guard, along with the other services, began a massive and rapid demobilization. Consequently, the Coast Guard shrank from a wartime high of more than 170,000 officers and enlisted personnel to less than 20,000 by mid-1947. Ocean stations, small-boat stations, and other shoreside units were closed. The sixteen *Tacoma*-class frigates that had been turned over by the Navy were decommissioned by the Coast Guard by mid-September 1946 and returned to the Navy, with some of them having served as Coast Guard cutters for only ninety days, and some of the cutters in the *Owasco* class were laid up in the period 1948–1951.

Additionally, there was congressional resistance to appropriate funds for the Coast Guard to do anything more than literally guard the coasts of the United States. Thus, the appropriation for fiscal year 1948 (then beginning July 1, 1947) provided funding for operation of only three ocean stations.

Hence, in the Atlantic, the Coast Guard patrolled two ocean stations: Stations A and C. Then, in October 1947, the crew of *Bibb* rescued all sixty-nine passengers and crew from a ditched aircraft on Station C. Because of this rescue, Congress provided fiscal year 1949 funding (beginning July 1, 1948) to the Department of the Treasury for the Coast Guard to operate all the proposed U.S. ocean stations, thereby enabling the service to meet the nation's obligation.[41]

Thus, the laid-up *Owasco*-class cutters began to be recommissioned in late 1948, the Coast Guard obtained more *Casco*-class ships on loan from the Navy, crews were found to man these cutters, and ocean station patrols proceeded in earnest.

As required by the 1946 agreement, the Second Conference on North Atlantic Ocean Stations met in London in April–May 1949, and the conferees included the nine original signatory nations of the 1946 agreement plus Denmark and Portugal. Though the conferees did not dispute the benefit of the ocean stations to international aviation and commerce, the nations were still emerging from an expensive war and were reluctant, if not hard-pressed, to come up with the money necessary to pay the crews, maintain the ships, and buy the fuel to support thirteen stations. And of course there was the inevitable question of how much each country should contribute both financially and operationally.

The result was a North Atlantic Ocean Weather Stations Agreement of 1949 with a reduction in the number of stations from thirteen to ten (via closure of Stations F, G, and L) for a new three-year operating period beginning July 1, 1950.

The operation of one station would be shared by The Netherlands and the United States (Station A); the operation of one station would be shared by Canada and the United States (Station B; Canada's obligation would be fulfilled via a bilateral agreement); four stations would be operated by the United States (Stations C, D, E, and H); two stations would be operated by the United Kingdom (Stations I and J); one station would be operated by France (Station K); and one station would be operated by Norway (Station M).[42] The participants at the second conference also agreed to another review conference in 1952 before the operating period of the agreement came to its end.

In May 1952, the eleven signatory nations of the 1949 agreement approved a protocol extending the terms of the 1949 agreement for a period of one year and postponing the review conference for a similar period.

In July 1953, the conferees at the Third Conference on North Atlantic Ocean Stations, held in Brighton, England, could only agree that a fourth conference was needed. In the end, the Fourth Conference on North Atlantic Ocean Stations, held in Paris, France, in February 1954, set the stage for the network of ocean stations for the next two decades. The thirteen signatory nations of the Agreement on

North Atlantic Ocean Stations of 1954 were Belgium, Canada, Denmark, France, Ireland, Israel, Italy, Norway, Sweden, Switzerland, The Netherlands, the United Kingdom, and the United States.

Because finances again were an issue, the conferees decided to reduce the number of stations from ten to nine; thus, Station H, off the East Coast of the United States at position 36°00' N, 70°00' W, was closed. (In 1970, Station H would be reactivated by Congress to monitor severe storm systems and manned by the Coast Guard at position 38°00' N, 71°00' W.)

The operation of one station would be shared by Norway and Sweden (Station A); four stations would be operated by the United States (Stations B, C, D, and E; an obligation of Canada to share the operation of Station B was fulfilled via a bilateral agreement); two stations would be operated by the United Kingdom (Stations I and J); one station would be operated by France (Station K); and one station would be operated by The Netherlands (Station M). The nations that did not provide ships and crews would make cash reimbursements or other contributions to balance their benefits (see Chart 2).[43]

That took care of the Atlantic. The Pacific, however, was another matter.

AT THE END OF World War II, the Navy retained operational control of the Atlantic and Pacific stations and began to reduce the number of stations. When the Coast Guard assumed operational control of the stations on July 1, 1946, the Pacific had five active stations. The Coast Guard immediately reduced that to two stations—Able, later known as Station P; and Fox, later known as Station N—explaining that budgetary constraints prohibited the service from operating any more than two stations in the Pacific. Later, Able would temporarily be closed during fiscal year 1948.

On July 25, 1949 (about 2½ months after the conclusion of the Second Conference on North Atlantic Ocean Stations), representatives of Canada and the United States met in Washington, D.C., to discuss establishment of a Pacific Ocean Station Program.[44] The

United States agreed to operate Stations N (30°00' N, 140°00' W) and O (40°00' N, 142°00' W) and to establish Stations Q (43°00' N, 167°00' W), S (48°00' N, 162°00' E), and T (14°00' N, 133°00' E) when funds for the latter three stations became available. Canada agreed to take on full-time operation of Station P (50°00' N, 145°00' W) in the Pacific in exchange for the United States taking over full-time operation of Station B in the Atlantic. Finally, it was assumed by both nations that Japan would continue its operation of Station X (39°00' N, 153°00' E), which had been established by the Japanese in 1947. This bilateral agreement came into force in June 1950.

After the onset of the Korean War in June 1950, Station S was activated in September 1950. The bilateral agreement was modified in February 1951 by request of the United States "in order that urgent requirements of an emergency nature may be met in the Pacific area." With the concurrence of Canada, Stations N, S, and P remained active; Station Q was activated; Station O was closed; and Stations U (26°00' N, 149°00' W) and V (33°00' N, 164°00' E) were established and activated, these two new stations operated by the United States. The two nations again assumed that Japan would continue to operate Station X.[45] The Japanese would conduct meteorologic and oceanographic research on Station X until late 1954.

The proposed Station T was ultimately eliminated from further consideration or action; however, Station T had originally been established by Japan in 1948 at position 29°00' N, 135°00' E, and was manned each year by the Japanese during typhoon season (i.e., June through October), which operation would continue until 1981.[46]

The following was noted in the modified U.S.–Canadian agreement of February 1951: "Upon the termination of the emergency, the program of Pacific Ocean Stations to meet peacetime requirements will be resumed."

With the end of the Korean War in July 1953, the number of Pacific stations began to be reduced, so that by 1955, the United States operated only two Pacific stations, N and V, and Canada continued its operation of Station P (see Chart 2).

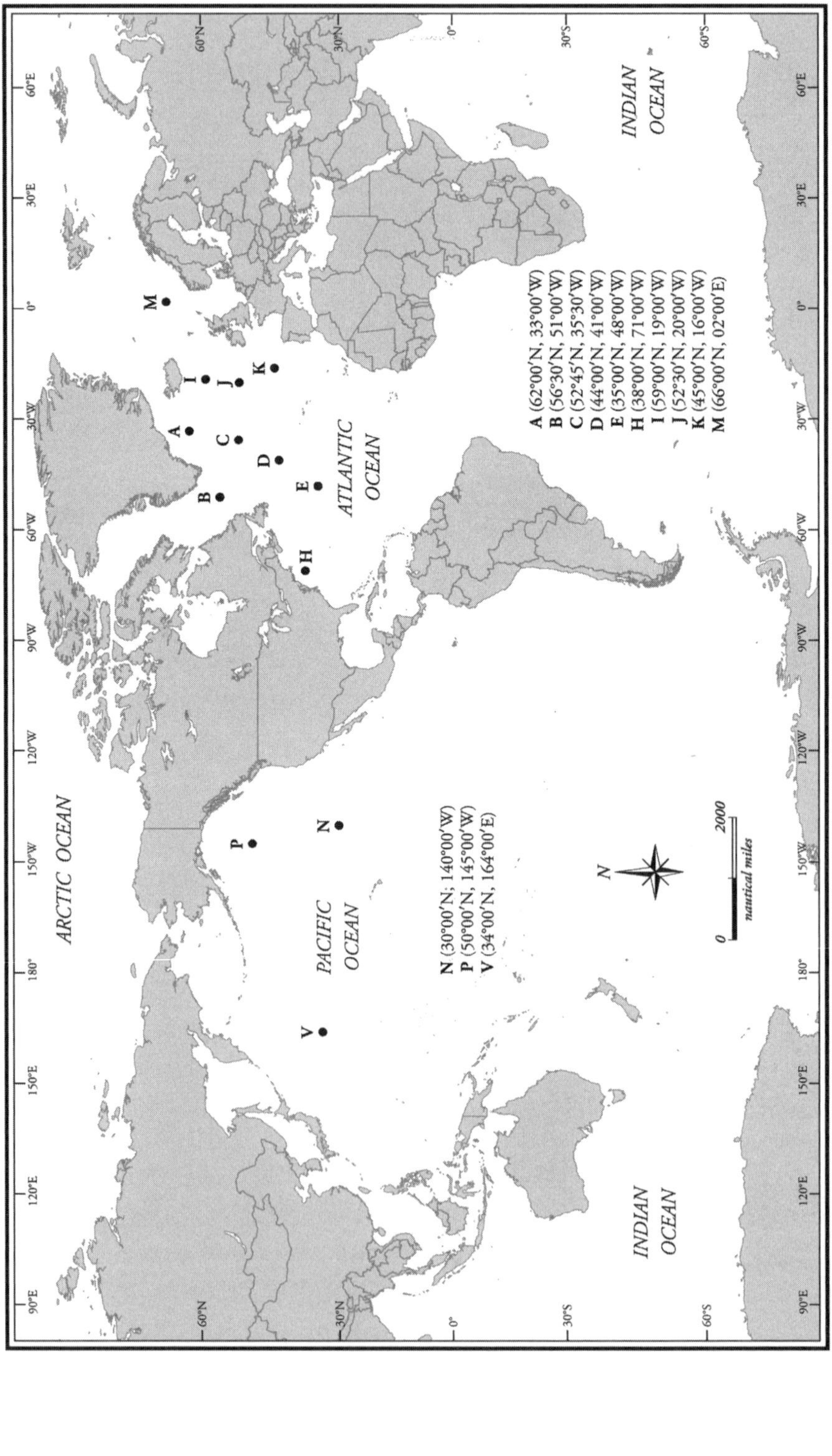
ARCTIC OCEAN
PACIFIC OCEAN
ATLANTIC OCEAN
INDIAN OCEAN
INDIAN OCEAN
A (62°00'N, 33°00'W)
B (56°30'N, 51°00'W)
C (52°45'N, 35°30'W)
D (44°00'N, 41°00'W)
E (35°00'N, 48°00'W)
H (38°00'N, 71°00'W)
I (59°00'N, 19°00'W)
J (52°30'N, 20°00'W)
K (45°00'N, 16°00'W)
M (66°00'N, 02°00'E)
N (30°00'N; 140°00'W)
P (50°00'N, 145°00'W)
V (34°00'N, 164°00'E)
0
2000
nautical miles
N

CHART 2. OCEAN STATIONS AFTER 1954. *Atlantic:* Shown are the positions of the nine North Atlantic ocean stations reestablished in 1954 at the Fourth Conference on North Atlantic Ocean Stations. Shown also is the 1970–1977 position of Station H (38°00' N, 71°00' W), which was closed as an ICAO agreement station (at position 36°00' N, 70°00' W) in 1954 but reactivated in 1970 by the U.S. Congress. Station A (62°00' N, 33°00' W) was operated by Norway and Sweden; Stations B (56°30' N, 51°00' W), C (52°45' N, 35°30' W), D (44°00' N, 41°00' W), and E (35°00' N, 48°00' W) were operated by the United States; Stations I (59°00' N, 19°00' W) and J (52°30' N, 20°00' W) were operated by the United Kingdom; Station K (45°00' N, 16°00' W) was operated by France; and Station M (66°00' N, 02°00' E) was operated by The Netherlands. Per the Agreement on North Atlantic Ocean Stations of 1954, Station M would later be operated by Norway and Sweden, and the operations of Stations A, I, J, and K would be shared by France, The Netherlands, and the United Kingdom. (Source for station positions: United Nations, *Treaty Series,* No. 2922, Vol. 215, p. 268. Source for reestablished Station H position: U.S. Coast Guard Commandant, 1970.) The Agreement on North Atlantic Ocean Stations of 1954 was terminated on June 30, 1975.[47]

Pacific: Shown are the positions of the Pacific ocean stations established through bilateral agreement between the United States and Canada and in operation after 1954. Stations N (30°00' N, 140°00' W) and V (34°00' N, 164°00' E) were operated by the United States. Station P (50°00' N, 145°00' W) was operated by Canada. (Source for positions of Stations N and P: United Nations, *Treaty Series*, No. 900, Vol. 70, p. 115. Station V was introduced in 1951 at position 33°00' N, 164°00' E, and its position moved to 34°00' N, 164°00' E, in March 1955. Sources for positions of Station V: United Nations, *Treaty Series*, No. 900, Vol. 87, p. 390; U.S. Coast Guard Commandant, 1972.[48])

Chart prepared by International Mapping.

ENDNOTES

1. Rohwer (1983, p. 56).

2. U.S. Coast Guard Cutter *Duane* (1941).

3. U.S. Coast Guard Headquarters Public Information Division (1949b, 1949c); U.S. Coast Guard Historian's Office (ship's files for *Muskeget* and *Monomoy*).

4. Morison (1957, pp. 96–97, chart).

5. Rohwer (2005, p. 194).

6. Williams (1963, p. 148). See Williams for the biography of Maury.

7. The charts developed by Maury and his staff and published by the Naval Observatory were catalogued as Series A to F: Series A, Track Charts; Series B, Trade Wind Charts (including monsoons for the Indian Ocean); Series C, Pilot Charts; Series D, Thermal Charts; Series E, Storm and Rain Charts; Series F, Whale Charts (Williams, 1963, p. 187).

8. Williams (1963, p. 187). The Storm and Rain Charts series was ultimately incomplete: Charts for the South Pacific Ocean and the Indian Ocean were never issued, likely because of the Civil War (p. 532, fn. 42).

9. In 1906, the U.S. Weather Bureau began a project involving fifty ships whose crews would "take daily weather observations at a fixed hour when in prescribed ocean locations" and radio the information to the Weather Bureau in Washington, D.C. (Willoughby, 1989, p. 127).

10. See Appendix I.

11. Wilbur (1969 reprint). Three Navy flying boats—NC-1, NC-3, and NC-4—together made the trans-Atlantic attempt, but NC-1 and NC-3 were forced to make water landings near the Azores and did not complete the flight.

12. Kauffman (1995, p. 188).

13. Subsequent devices developed for early aircraft radio navigation include the Aircraft Radio Corporation (ARC) navigation receiver and the Sperry Gyroscope Company automatic direction finder (ADF). For photographs and brief descriptions of the original Leuteritz, ARC, and Sperry devices, see the National Air and Space Museum Web exhibit "America by Air" at: http:/

/www.nasm.si.edu/americabyair/innovation/innovation06.cfm. Accessed 1 June 2009.

14. MacTaggart-Cowan (1938); Lawrence (2004, p. 120).

15. See Appendix I.

"I believe that a northern transatlantic regular service should not be contemplated with planes which are not capable of flying nonstop from the western side of the Greenland ice cap to Iceland if necessary. . . . It is not possible for me to overemphasize the necessity of a sufficient number of radio stations on a northern route to give reliable weather information and to give bearings." From a letter of Charles A. Lindbergh to Juan T. Trippe, September 15, 1933 (Pan American World Airways, Inc. Records, Special Collections Division, University of Miami Libraries). In May 1927, Charles "Lucky Lindy" Lindbergh made the first solo nonstop trans-Atlantic flight. He flew from New York to Paris in the *Spirit of St. Louis* using dead reckoning and celestial navigation, and he later served as the technical advisor to Pan Am.

16. See Appendix I.

17. The Grand Banks Patrol was a facet of the pre-war Neutrality Patrol.

18. See Appendix II.

19. See Appendix I.

20. The Coast Guard was returned to the Department of the Treasury on January 1, 1946, by Executive Order 9666 of December 28, 1945.

21. For the involvement of the Coast Guard with BOLERO, see footnotes 22, 23, and 27. For details on BOLERO and POINTBLANK, see *United States Naval Administration in World War II. United States Naval Forces, Europe. Volume V. The Invasion of Normandy: Operation NEPTUNE.* London: Commander, U.S. Naval Forces in Europe, 1946. Mss. #147-E. Available at the Naval History & Heritage Command website: http://www.history.navy.mil/library/online/comnaveu/comnaveu_index.htm. Accessed 1 June 2009.

22. U.S. Coast Guard Headquarters Public Information Division (1949b). Available Coast Guard historical sources do not specify either the coordinates of the first two plane guard stations (only the geographic locations) or the exact dates of establishment of these two stations (only "summer 1942"). However, coordinates for the stations to be manned by the Greenland Patrol

are provided in a memo (i.e., operations orders), "Bolero Ferry Plan," prepared on behalf of the commanding general by Col. John K. Gerhart, AAF (S-3; Headquarters Eighth Fighter Command, Grenier Field, Manchester, New Hampshire), dated June 18, 1942. The coordinates of the plane guard stations given in the memo (A: 60°00' N, 53°00' W; B: 64°00' N, 30°00' W) comport with the general geographic locations of the stations noted in U.S. Coast Guard Headquarters Public Information Division (1949b). A researcher located a working copy of this memo and other documents related to the Bolero Ferry Plan at the National Archives (College Park, Maryland). To view a copy of the Gerhart memo, see the link provided at http://www.noreasterpress.com/documents/Bolero_links.php.

Of note, the Gerhart memo states that in addition to the cutters manning Stations A and B, two Greenland Patrol cutters would be stationed off the western coast of Greenland, one each at the entrance to "BW-1 fjord" (the fjord leading to Bluie West 1 air base) and at the entrance to "BW-8 fjord" (the fjord leading to Bluie West 8 air base) to identify the precipitous entrances for the aircraft and to make weather observations.

"In Denmark Strait and in Davis Strait, to the east and west of Greenland, and in the fjords leading to BW-1 and BW-8 were Coast Guard cutters to act as navigation markers" (Milner, 1947, p. 219). According to Milner's citations, the June 18, 1942, memo with annexes appears in *Eighth Air Force History*, I. See Milner for other source documents. The ferry flights began on June 23, 1942. See also footnotes 23 and 27.

23. The proposed technique by which cutters on plane guard stations would serve as navigational waypoints for BOLERO aircraft is described in Part Two of a May 24, 1942, encoded Telex sent by Commander Task Group 24.8, SOPA Greenland (COMGREENPAT; Capt. Edward H. "Iceberg" Smith): "On twenty nine May occupy above stations [see footnote 27] except Denmark Strait and hold communication test with Army bases and Five Baker Seventeen on weather reconnaissance. Primary frequencies 355 and 4050. Secondary 2670. For emergency homing 414 kcs. Plane guard vessels take bearings of planes 355 transmissions plotting approximate plane positions." To view a copy of Parts One and Two of the decoded Telex, see the link provided at http://www.noreasterpress.com/documents/Bolero_links.php.

24. Dinsmore provides the April 1, 1944, date based on station reorganization by COMTASKFOR 24. See Dinsmore, R.P., Captain, USCG (Ret.), "Ocean Weather Ships 1940–1980." U.S. Coast Guard Historian's Office, Washington, D.C. Available at: http://www.uscg.mil/history/webcutters/

rpdinsmore_oceanstations.asp. Accessed 1 June 2009. U.S. Coast Guard Headquarters Public Information Division (1949b) provides a time frame of reorganization "by June 1944" when the plane guard reporting system was administratively incorporated into the weather observation reporting system. Willoughby (1989, pp. 108, 128) provides the fall 1942 transfer date and notes that the expansion of four stations occurred in spring 1944. Though the stations were reorganized, the duties of plane guard ships and of weather observation ships were not made uniform until after V-E Day (U.S. Coast Guard Headquarters Public Information Division, 1949b).

25. See Appendix III.

26. U.S. Coast Guard Headquarters Public Information Division (1949a); U.S. Coast Guard Historian's Office (ship's files for *Escanaba*).

In 1946, the service would name one of its new *Owasco*-class cutters *Escanaba*, and thus a second cutter with that name would patrol ocean stations into the 1970s. That *Escanaba* was decommissioned in 1974, and the service commissioned a third *Escanaba* (*Famous* class; 1,800 tons, 270 feet) in 1987.

27. According to Part One of the May 24, 1942, encoded Telex sent by Captain Smith: "US Air Force Ferry Command plans initiate continual 'Bolero Movement' from Goose Bay through Greenland bases. Six steel trawlers being converted as plane guard vessels for Greenland Patrol one each with 60 mile radius at Station A 60-00 North 53-00 West and Station B 64-00 North 30-00 West. One each at entrances of BW1 and BW8 fjords." The coordinates of the stations comport with those provided in the Gerhart memo. Cutters on Stations A and B were required to keep station within a 60-mile radius of the station coordinates. Which trawlers besides *Aivik* patrolled the stations could not be determined. A May 25, 1942, encoded Telex sent by COMTASKFOR 24 affirms: ". . . Greenland Patrol provide weather broadcast and flight security ships at (1) entrance Bluie West One, (2) entrance Bluie West Eight, (3) Davis Strait, (4) Denmark Strait. . . . Location of ships in Davis Strait Strait [sic] and Denmark Strait and communication provisions as arranged with Army subordinate ferry command by Commander Greenland Patrol." To view a copy of the decoded May 25 Telex, see the link provided at http://www.noreasterpress.com/documents/Bolero_links.php. According to a citation in Milner (1947), the May 25 communiqué of COMTASKFOR 24 is located in Bolero File, AAF Archives.

28. Johnson, Robert Irwin. "Coast Guard–Manned Naval Vessels in World War II." U.S. Coast Guard Historian's Office, Washington, D.C. Available

at: http://www.uscg.mil/History/articles/h_cgnvy.asp. Accessed 1 June 2009.

29. USS *Big Horn* (AO 198), which became USCGC *Big Horn* (WAO-124), is not the current U.S. Military Sealift Command vessel USNS *Big Horn* (T-AO 198).

30. *Sea Cloud* plies the seas today as a sailing cruise ship.

31. Pierce et al. (1948, p. 19).

32. U.S. Coast Guard Headquarters, Office of Engineering (1944, pp. 4, 5).

33. Pierce et al. (1948, p. 20).

34. U.S. Coast Guard Headquarters, Office of Engineering (1944, p. 9); Pierce et al. (1948, p. 24). The term *loran* (long range radio navigation or long range navigation) was coined by Commander Harding in summer 1942.

Similar to the term *radar* (radio detecting and ranging), reference to the technology is traditionally lowercase (loran), whereas reference to a particular mode is capitalized (e.g., Loran, Loran-A, Loran-C) (Hefley, 1972, p. xi).

35. "During World War I, a method was developed for locating enemy guns by measuring the differences between the times at which their reports were heard at three different listening posts. The posts were electrically connected so that the three detonations received from a shot from a given gun could be recorded on one chronograph. The time interval, from the arrival of the report at one post to its arrival at another post, is a measure of the amount by which the gun's distance from the latter post exceeds its distance from the former. This difference is precisely the constant that defines a hyperbola with respect to the two posts as foci. The time interval between the reception of the gun's report at the middle listening post and its reception at one end post locates the gun upon a hyperbolic line of position that passes between the two posts and is concave toward the nearer one. The interval between the received reports at the middle post and at the other end post locates the gun upon another hyperbolic line of position. The two hyperbolas intersect at the gun." (Pierce et al., 1948, p. 10.)

In loran, the principle of the system is the same, but the direction of travel of the signals is reversed: The "three listening posts" are instead three shore-based transmitters issuing time-coordinated radio pulses, with each transmitter emitting its own distinctive pulse; the "gun" is instead the receiver, used by the navigator to determine his own position in relation to the three transmitters. In loran chains, the arrangement of a single master station (the "middle listening post") working in tandem with two slave stations (in effect, two master–slave pairs of stations) is known as a *triplet.*

36. U.S. Coast Guard Headquarters, Office of Engineering (1944, p. 16).

37. U.S. Coast Guard Headquarters, Office of Engineering (1944, pp. 6, 13); Pierce et al. (1948, p. 28).

Operational use of loran began in the U.S. Navy in spring 1943, followed by use in the Royal Canadian Navy and the Royal Navy. A reliable airborne model of loran receiver-indicator became available in summer 1944, and loran was then used by USAAF and USN, Royal Canadian Air Force, and Royal Air Force planes in antisubmarine, reconnaissance, and convoy escort missions (Pierce et al., 1948, pp. 28–30).

38. In the Atlantic, by May 1945 sixteen stations had been established north of 15° N, and six stations had been established between the Equator and 15° N (Willoughby, 1989, p. 129). Of the sixteen stations located north of 15° N, eleven were U.S. stations manned by the Coast Guard (Stations 1 to 11), and five stations near Europe were manned by the United Kingdom. Of the six stations located south of 15° N, four were manned by Brazil, and two were manned by the United Kingdom. On November 1, 1945, the Coast Guard assumed two of the Brazilian stations, and by March 1, 1946, all stations south of 15° N were closed (U.S. Coast Guard Headquarters Public Information Division, 1949b).

The British operated several weather stations in the North Atlantic before the entry of the United States into World War II.

39. Provisional International Civil Aviation Organization (1946). The 1946 agreement, however, never entered into force (Wiktor, 1998, p. 505).

40. The locations of the thirteen ocean stations proposed by the International Agreement on North Atlantic Ocean Weather Stations (1946) were as follows: A (62°00' N, 33°00' W); B (56°30' N, 51°00' W); C (51°45' N, 35°30' W); D (45°00' N, 45°00' W); E (34°00' N, 52°00' W); F (35°30' N, 40°00' W); G (46°00' N, 29°00' W); H (36°00' N, 70°00' W); I (60°00' N, 20°00' W); J (53°50' N, 18°40' W); K (47°00' N, 15°00' W); L (39°00' N, 17°00' W); and M (66°00' N, 02°00' E) (Provisional International Civil Aviation Organization, 1946). Some station positions would be modified under the subsequent 1949 agreement.

41. Foley (1948).

42. United Nations, *Treaty Series*, No. 1405, Vol. 101, p. 91. The locations of the ten ocean stations established by the North Atlantic Ocean Weather Stations Agreement of 1949 were as follows: A (62°00' N, 33°00' W); B (56°30' N, 51°00' W); C (52°45' N, 35°30' W); D (44°00' N, 41°00' W); E (35°00' N, 48°00' W); H (36°00' N, 70°00' W); I (59°00' N, 19°00' W); J (52°30' N, 20°00' W); K (45°00' N, 16°00' W); and M (66°00' N, 02°00' E).

43. United Nations, *Treaty Series*, No. 2922, Vol. 215, p. 249, p. 268; Lister (1954). Per the agreement, Station M would later be operated by Norway and Sweden, and the operations of Stations A, I, J, and K would be shared by France, The Netherlands, and the United Kingdom.

44. United Nations, *Treaty Series*, No. 900, Vol. 70, p. 115.

45. United Nations, *Treaty Series*, No. 900, Vol. 87, p. 390.

46. IOC/WMO (1978, p. 10); Toba et al. (1983, p. 37).

47. Wiktor (1998, p. 596).

48. According to the March 1955 logbook of the cutter *Matagorda*, Station V was established at the new position at 1200 on March 12, 1955.

3

BRAVO

56°30′ N, 51°00′ W

THERE WAS A PEACETIME ENEMY on the ocean stations. This enemy was not always threatening but was always ready to strike. And when it struck, it struck with ferocity. Weather—the very weather that the cutter crews were supposed to observe and report—was the enemy.

Coast Guard cutters on ocean stations had their hulls fractured, their topside structures smashed, their boats destroyed, and their gear carried away by the sea. Although these cutters were comparatively small ships, they were not inherently fragile: They were built of steel to the standards of construction required for warships, and they were sailed by men who had experience, men who knew and understood the sea.

One such cutter was *McCulloch*, a 2,500-ton, 311-foot sister ship to *Absecon*. Both of these cutters were *Casco*-class vessels (former Navy seaplane tenders) that came to the Coast Guard in the latter half of the 1940s. Eventually, the Coast Guard would have eighteen of these vessels, all of them patrolling ocean stations.

McCulloch sailed to Ocean Station Bravo in late December 1958. Her captain, Cdr. Albert F. Wayne, was a veteran of ocean stations and of the sea. He had been across the Columbia River Bar, the West

Coast river entrance known for its huge, breaking swells. He had also been through hurricane conditions aboard various Coast Guard cutters on at least five occasions before his cutter arrived on station on December 28. Consequently, he knew that he had to have his crew make substantial preparations for a winter North Atlantic Ocean: The boats were all cradled, extra lifelines were rigged on deck, and all topside gear was lashed down.

The enemy, however, conspired against the cutter. A low-pressure center lay in wait for a week off the eastern coast of Canada. As it sat, unmoving, it built in intensity. *McCulloch* sat motionless, too, with comparatively mild weather until January 2, 1959. When the wind started to build from the northeast in the morning, it did so quickly, and likewise it quickly built seas of twenty to twenty-five feet. *McCulloch* had gotten under way at 0330 just to remain in the center of station, but as conditions deteriorated, the cutter stayed under way to keep heading into the seas.

By the next morning, January 3, the wind was blowing sixty knots, and the seas were running forty feet. Captain Wayne and *McCulloch* and her crew had endured such weather before, but this storm would not end in a day or two, nor would it pass by them causing the wind to shift. Instead, the low-pressure system just generated stronger and stronger winds from the same direction, which, because the center of the system did not move, generated higher and higher waves.

A SHIP IN A SEAWAY is subjected to continuous abuse. As each wave moves by the ship, it first picks up the bow, then the middle of the ship, and finally the stern. This motion puts nonstop stress on the hull. Imagine picking up a long, narrow, cardboard box by its ends: The middle of the box sags.[1] (Naval architects refer to this as sagging.) Now imagine the box picked up in the middle: The ends sag. (To the naval architect, this is hogging.) Pick the box up at opposite corners and it twists (wracking). In a heavy sea, a ship shifts between sagging and hogging and wracking and sagging over and over again. Do this enough times to a box (or even a piece of steel), and eventually it breaks.

As the ship pitches (first the bow rises and then drops as the stern rises) across the seas, hogging and sagging occur continuously. A ship, however, is a box with a fixed top on it, its main deck. This "top" usually provides enough rigidity to minimize the sagging and hogging and wracking so that the ship stays in one piece.

The seas, however, are neither of a uniform height nor do they come from a uniform direction nor are they a uniform distance apart. So the ship not only sags and hogs and wracks but also gets pushed off its intended course (yaws) because of waves striking slightly from one side then the other. These nondirect waves also cause the ship to roll from side to side.

As the stern rises during the pitching, it can rise far enough that the propeller comes partway out of the water. This causes uneven thrust, and thus the ship will slow down then surge ahead once the propeller regains a full bite. Simultaneously, the ship will surf down the "back edge" of each wave amplifying the surge.

The ship is also lifted bodily straight (or almost straight) up and then let down, a phenomenon known as heave. So a ship in a seaway is sagging, hogging, wracking, pitching, yawing, rolling, surging, and heaving. It is moving in all possible directions in all three dimensions with no true consistency to its motion.

Of course, the people inside the ship are subjected to all this motion as well. They walk as if intoxicated because the decks are constantly moving beneath them. If they do not use hand-holds, they are literally thrown about. Staying in bed is one way of avoiding the thumps and bumps and bruises, but sleep is impossible. The ship's motion will roll a crewmember out of his rack or pitch him up and down sufficiently so that he momentarily floats in the air above the mattress.

Everyone will become tired, and most will feel the disorientation that is caused by the motion. It is the unusual human being who can withstand all this without becoming seasick. Seasickness can leave its victim weak and dehydrated and can become life-threatening. It is misery and can punish the crew just as the sea punishes the ship.

As a wave slides beneath the ship, the bow rises. If the seas are big enough, the entire bow will shoot out of the water over the crest and hang in the air until gravity takes over and slams the bow back into the trough of the sea. Then, the crew hopes, the bow will start to rise to meet the next wave. If the waves are far enough apart, and there is enough buoyancy and enough freeboard (i.e., the vertical distance between the waterline and the deck) at the bow, then the ship will pitch up and over the next wave.

But not always. When the wind is blowing hard enough and the seas are close enough together, the waves will lose their smooth form and will begin to break, just as they do at a shoreline, and a whole wall of water will smash down on the ship's foredeck. This wall of water sometimes crashes down on decks well above the main deck, inundating the whole forward portion of the ship.

It is then that the ship feels as if it is a thing alive. As the tons of water wash around the deck structures and begin to cascade overboard, the entire ship will begin to shake. It feels like what a slow-motion movie of a dog shaking off water looks like: the vibration starts at the bow and slowly moves aft until it disappears somewhere astern.

The challenge for the ship, however, is that this is not a one-time occurrence. As soon as one wave crashes aboard, smothering everything in a wash of foam, another wave is preparing to do the same thing. This goes on, over and over, minute after minute, the ship shedding the water dozens of times an hour, hundreds of times in a four-hour watch, thousands of times as the storm passes.

IT IS NOT AMAZING that ships succumb to this violence and break or break apart; what is amazing is that most ships survive it. *McCulloch* continued to survive this weather, but she was taking a beating. On the morning of January 3, the wind was blowing at a steady sixty knots, near hurricane force, and by that afternoon there was a steady seventy-five knots of wind with gusts of ninety knots. Worse, it blew that hard out of the east or east-northeast, the same direction it had been coming from for more than thirty-six hours. Inevitably, the seas continued to build, and by the afternoon they were more than fifty feet high, and the crests were breaking continuously. The crew on the bridge had to look up to see the top of the next wave, and as they did so they had to stare at a solid wall of water that was about to bury their cutter's bow.

McCulloch shook off dozens of these seas, each one breaking aboard. Then her bow would lift once again, she would pitch over the crest, her bow coming completely out of the water until it would slam back down into the ocean by which time the next wave would begin to break and start the process over again.

At 1310, *McCulloch* was struck on the port bow by what Captain Wayne described as "one tremendous wave." As it was he who had estimated the average height of the seas as being fifty feet or more, this wave must have been truly terrible. Its effect was terrible: it demolished *McCulloch*'s forward jackstaff, carried away sections of the main-deck life rail, knocked loose a ready-box (a steel cube about four feet on each side for holding ammunition) that was welded to the 01-deck (the deck *above* the main deck), ripped one life raft free from its rack, bashed in a steel bulkhead on the main deck, and destroyed the hedgehog mount (a large, rocket-launched antisubmarine weapon) also located on the 01-deck.

Sailors had for years reported "monster waves" that seemed to come from out of the middle of such storms, bash their ships, and then recede into the distance. Such reports increased as stronger ships survived the same types and sizes of waves that had undoubtedly sunk smaller vessels in centuries past. The scientific community was long reluctant to accept the existence of such freak, or rogue, waves because of a lack of quantifiable and verifiable records. (The reports of sailors were considered "sea stories.") However, improved oceanographic analysis and modeling, satellite-based radar imagery, sensors on deep-sea rigs, as well as an increased history of ships encountering and surviving these waves has confirmed their existence.

The wave that struck *McCulloch* was not finished, however, because it washed the ready-box all over the forward part of the cutter, knocking down ladders, destroying life-raft racks, and smashing against all the doors, rails, fittings, and bulkheads in the forward part of the cutter. Because there were still more fifty-foot seas coming aboard, this box would become a weapon threatening the very cutter itself: If it poked a hole in the superstructure or deck, water would begin to enter uncontrollably. The cutter could be sunk. Some of *McCulloch*'s crew had to go on deck to lash it down.

It was cold. The air temperature was 36°F and the water temperature was the same. Anyone going outside to work would want warm clothing, and—in those days before widespread synthetic materials—that meant wool sweaters, denim trousers, and wool or cotton gloves. It was raining: Because of that and the wind-driven spray and the breaking seas, the deck was wet and slick. Anyone going outside to work would want waterproof clothing and that meant raincoats and rain pants and boots. The men dressed in the heavy clothing, trundled out, wrestled the ready-box into submission, and lashed it down.

Then tragedy struck. Seaman John H. Lewis, Jr., was swept overboard. The rain-coated decks; the near-freezing temperatures, which numbed fingers and feet; the cumbersome clothing; and the pitching, rolling, heaving, surging cutter nearly guaranteed someone would fall down. And once down, any man would have been hard pressed even to crawl back to safety, let alone regain footing before another lurch of the cutter or rush of water would have pushed him across the deck and under the lifeline and into the winter North Atlantic.

McCulloch's crewmen struggled to recover their lost shipmate, but within a few minutes he was gone.

SOMETIME IN THE MIDDLE of this maelstrom, *McCulloch*'s crew noticed that the interior main deck covering (a substance generically referred to as "underlayment," which served as both insulation and a smooth surface over which tile or linoleum could be laid) was buckling and breaking up. They feared that the main deck itself was cracking.

It was.

If the main deck cracked significantly, the whole structure of the cutter would be threatened. Small cracks could spread, join each other, and the cutter could be opened up across its entire width. Then, the rigidity provided by the four-sided "box" of the cutter's structure could be lost. Or the cracks could spread from the deck to the sides of the hull with an even greater loss of rigidity.

One crewman was dead, there was gross damage to the cutter's doors, bulkheads, lifelines, and equipment, and now the cutter was, literally, breaking up. *McCulloch* was in dire straits.

On the morning of January 4, twenty hours after Seaman Lewis was swept overboard, the wind, still out of the east, began to ease slightly "down" to about fifty knots. The seas lay down, running at "only" thirty-five feet. Captain Wayne decided it was time to turn around in an attempt to move away from the worst of the storm. This would be no easy task. If his timing was off, the cutter could roll precipitously, allowing water to wash across the decks and down ventilators, possibly sinking the cutter. If another of those "tremendous" waves suddenly appeared halfway through the turn, *McCulloch* could be capsized. It would take precision, a strong sense of the conditions, and positive action to make the evolution work.

Captain Wayne chose his time and made his move. Before he started the turn, he would have looked well ahead to ensure no abnormally high waves were in the immediate future. He could accelerate the rate of turn, thus minimizing the amount of time his cutter was broadside to the seas, by increasing speed on one shaft and decreasing it on the other shaft as well as using the rudder. He would want to initiate the turn when he was reasonably certain the rudder and propellers were fully immersed and not when the ship pitched down the face of a wave with its stern up in the air. It is an evolution that has no textbook answer but demands experience to be successfully executed.

The cutter rolled, but never more than forty-five degrees. *McCulloch* got on her new course, still upright, but now going downwind and down swell. She would survive. On the morning of January 5 she reached the edge of Ocean Station Bravo, still heading west while the winds continued to blow from the east at nearly forty knots. The next day she reached the shelter of Hamilton Inlet, Labrador.

Later that afternoon, those ashore who directed *McCulloch*'s operations (Commander Eastern Area)[2] decided that enough was enough. *McCulloch* was told to abandon station and head home for Boston. Her damage would be evaluated and repairs started there.

The repair list had more than forty items on it: eight cracks in the main deck, several fractured welds on the underside of the main deck, main deck bulkhead dished in, port and starboard main deck doors and frames bent, rails missing, an overhead seam on the 01-deck open and leaking, life-raft racks damaged beyond repair, radio antenna bent, fuel oil tank bulkhead cracked, and ladders bent or destroyed. Nonetheless, *McCulloch* was repaired and returned to patrol the ocean stations for another dozen years.

MCCULLOCH'S JANUARY 1959 Bravo patrol was miserable and extreme, but not uncommon. The simple fact is the weather recorded and forecast for the ocean stations was often far less than idyllic. The northernmost stations frequently had extended periods of adverse weather. Crews would conduct entire patrols without ever once seeing the sun because of continual overcast. Intense low-pressure systems would sweep out from the cold reaches of Canada and pummel the cutters on these stations. As soon as one of these lows had passed, another would be in the brewing, and just as a cutter began to recover from its beating, a new storm would be on the way.

When an intense storm would pass a station, a cutter would have no choice other than the one made by *McCulloch*'s captain: head into the sea and swell for simple survival. Days of such heaving to would often result in the cutter slowly creeping ahead and thus being at the outer extremities of the grid by the time the storm had passed.

Sometimes, a cutter could be making turns (i.e., ordering the propellers to make speed) for five or six knots, but the wind would hold the cutter back, and the breaking seas would sometimes bring it to a shuddering halt. It was even possible for a cutter to move backwards, along the reciprocal of her ordered course, as she slugged her way through such seas for hours and sometimes days on end.

Usually, the storm would move along quickly enough so that the crew could head the cutter back toward the center of station before the cutter exited the far reaches of the grid. Often, a cutter crew would have just enough time to square itself away and return to the center of station before another storm would blast onto the station.

Rarely would a station be abandoned. The cutters had proved themselves tough by weathering many such storms, and an ocean station crew would remain on station when the crews of most other ships would make for safer waters.

As bad as *McCulloch*'s Bravo had been, Captain Wayne noted correctly that she might have been saved from worse catastrophe by virtue of the fact that the temperatures stayed just above freezing. Often during the winter months, air temperatures would be well below freezing while water temperatures hovered near the same mark. Wind-whipped spray as well as the water that climbed aboard a cutter's decks while it labored in a storm would quickly freeze when it struck the cold decks and bulkheads. As the ice built up, the ship became top-heavy, and without constant efforts at de-icing, capsize was possible.

This phenomenon was not new to the ocean station cutters nor has technology done anything to abate it in the years since the ocean station patrols ended. There are many well-known cases of vessels, notably crab boats off Alaska, that have iced up, lost stability, and capsized.

There were two simple devices that all the cutter crews found best for battling ice: wooden baseball bats and ax handles. These tools worked best because they were relatively light weight, designed to be swung against solid objects, and though hard enough to shatter the ice, they were soft enough not to damage the steel and wire that they struck. Crewmen buttoned up in their foul-weather gear would go on deck with the bats and literally beat away the ice. There was no better or more efficient method of ice removal than the hand-held bat, and no ocean station cutter ever headed for a northern station without two dozen bats stored in the boatswain's hold.

It would be just such icing that would cause another cutter to abandon Ocean Station Bravo only three years after *McCulloch*'s brutal patrol. The month was once again January, the year was 1962, and the cutter was the first-in-class *Owasco.*

OWASCO WAS DRIFTING in position OS on the morning of January 24, 1962. At 0400, she got under way as the wind quickly built to near-hurricane force (sixty knots) and the seas began to build. Worse, the air temperature, which had been at freezing, began to plummet into the twenties. The seawater temperature was 35°F. As the wind blew the spray against the cutter, it froze almost instantaneously.

That evening, "extremely heavy sea" was recorded in *Owasco*'s log. As had been true for *McCulloch*, this single wave wreaked havoc. One of the starboard-side boats, in its cradle on the 01-deck, some fifteen to twenty feet above the waterline, had both its bow and stern torn away. (The ocean station cutters carried as many as four small wooden boats, usually a combination of pulling boats and motor boats, until the middle of the 1960s. By then, the reliability of small diesel engines and reduced maintenance for fiberglass hulls had made the old, heavy, wooden boats unnecessary, and they were all removed from the cutters.)

On *Owasco*'s port side, the same wave lifted another boat out of its cradle and jammed it between the cutter's hull and a boat boom that was attached to the cutter's hull. Lifelines and stanchions on the foredeck were smashed, and some were carried away. The strike from this wave was so violent that one of the cutter's young officers "sustained back injuries as a result of the shock" according to the cutter's logbook.

The wind blew at seventy knots all night, and *Owasco* was bashed by one sea after another all night. The next morning, the wind began to diminish, but the damage was not over. The cutter's air-search

Owasco's damaged starboard pulling boat.

radar antenna, a wide and complex metal web, could no longer hold on against the wind and the violent motions of the cutter. It tore loose from its mounting and wedged itself against the mast. Now the threat that it might fracture completely and fall hung (quite literally) over everyone's head.

Nonetheless, things were improving because the winds and seas diminished until 2000 when the winds shifted to the east at twenty-five knots and the seas started to build from the new direction. Things were not improving with the temperature, which had dipped into the twenties the day before and now dropped into the teens. It would climb slowly back up to freezing and stay there throughout the following day, the 26th of January.

The wind remained at twenty-five knots but veered from east to south to southwest to northwest, and thus no truly big sea or swell could build. By 2000 on January 26, *Owasco* was in another storm with northwest winds blowing sixty knots, the seas building, and the cutter encountering a plunging, crashing, soaking ride. But the temperature continued to hover around freezing and thus *Owasco* avoided severe icing.

The cycle started over again on January 27 with the wind diminishing, shifting to the east and then southeast, and then increasing to forty-five knots. This day, though, the air temperature dropped below freezing and stayed there. Once again, the wind veered and increased to fifty knots. Once again, *Owasco* steered into the seas. Once again, the cutter's motion became so violent that a crewmember was thrown down and injured.

Then things got nasty. *Owasco* suffered through twenty-four hours of fifty-knot winds, which meant seas running thirty feet or more and a bow-burying ride that shot spray all over the cutter, from the main deck to the tops of the masts. The air temperature plunged into the single digits. She began to build ice and build it heavily.

The hours of daylight are short in the northern latitudes in winter. It would have been dangerous to send anyone on deck to break ice in the huge seas and foolish to have done so in the same conditions in the dark. *Owasco* plunged ahead, beginning to wallow because of the added topside weight, unable to do anything to ease her plight.

Owasco's air-search radar antenna wedged against the mast.

With daylight on January 29, Cdr. David L. Davies, *Owasco*'s captain, could no longer tolerate his cutter's deteriorating stability. He turned out all hands to break ice. The radar antenna still hung over their heads. The lack of lifelines and stanchions forward made it entirely too risky to venture out on the bow to break ice. The crew

Owasco's superstructure looking aft.

flailed away as best they could with their baseball bats, beating the ice off the superstructure, the remaining lifelines, and the various pieces of deck equipment.

By now, *Owasco*'s maneuvering through this series of storms had carried her off station. Her crew was beat and battered, half her boats were rendered useless, the radar antenna was still poised to wreak more havoc on deck, her stability though improved was still impaired, and the weather on Bravo in January was not going to get better. The Coast Guard's Commander Eastern Area directed *Owasco* to leave station and seek warmer temperatures and shelter. Three days later, she arrived in Argentia, Newfoundland, with the antenna still wedged in place.

To the men who went on ocean stations in the North Atlantic, *McCulloch*'s and *Owasco*'s January Bravo patrols were a yardstick by which endurance was measured. Any conditions less than those were, *de facto*, insufficient cause for abandoning station.

An *Owasco* crewman wields his ice-removal tool.

* * *

OWASCO AND HER SISTERS were all designed for the Coast Guard during World War II and built in the period 1944–1946. Two of the *Owasco*-class cutters (*Mendota* and *Pontchartrain*) were built by the Coast Guard itself at the Coast Guard Yard. These cutters displaced 1,980 tons and were 255 feet in length. When built, they carried incredibly heavy armament for such small vessels. *Owasco*, for instance, was built with two 5-inch/38-caliber twin-mounts (i.e., there were four 5-inch guns onboard), two 40-mm quad mounts, two depth charge racks, and six K-gun depth charge projectors. This was as much weaponry as would be carried on the most heavily armed of the *Secretary*-class cutters (which had 15 percent more displacement than the *Owasco*-class cutters) even at the height of World War II.

There was a persistent legend in the Coast Guard that the *Owasco*-class cutters, having been built during wartime, were originally designed to be a hundred feet longer than their final length, but a whole section of the vessel was simply deleted during the final planning/construction stages. The reasons for this deletion (according to the legend) are these: (1) a lot of steel was saved, thereby allowing construction of a greater number of cutters; and (2) as the "deleted" section included one entire engine room, machinery was likewise saved and used on other cutters.

It is easy to see how this legend came into being. The *Owasco*-class cutters did have only one engine room with a steam turbine driving a generator that drove an electric motor that drove a single shaft. It was a complex propulsion system, and it lacked the redundancy that was (and is) the norm for warships. It seems odd that a single-propeller small combatant would be designed during World War II when all warships were at great risk of battle damage. The twin-screw concept was well established and accepted for warships during that era, and the vulnerability of an engineering plant with a single turbogenerator–motor and single propulsion shaft raises a question about the wisdom of such a system.

Furthermore, it was difficult to get around inside the *Owasco*-class cutters. There seemed to be an interminable number of bulkheads with no doors belowdecks so that everyone had to go up and over to get from one compartment to the one adjacent. Certainly, these bulkheads were there to ensure the vessel's integrity in case of damage, but the lack of doors and the apparent "cut-up" nature of the internal design contributed to the "shortened-ship" legend.

The cutters were also "cut-up" topside with a divided superstructure. This idiosyncrasy was caused by the fact that the cutters were supposed to accommodate an aircraft amidships, but that requirement disappeared before any of the cutters were placed in service.

Next, the legend invited this legitimate question: Why did such small vessels have such huge gun inventories? The legend's answer: The cutters were designed to fight a war and therefore no one was going to reduce armament even though the size shrank.

Owasco class: The cutter *Klamath*.

The ultimate and most visible argument supporting the legend is that the *Owasco*-class cutters had an appearance that can only be described as ungainly. If one looks at these cutters with the foregone belief that they had an entire section (i.e., the "missing" engine room) deleted, the appearance of the cutters certainly serves the legend. They are stubby and very unlike the long, lean, *Secretary*-class cutters or the *Casco*-class cutters that patrolled with them on the ocean stations. They appeared to have abnormally high superstructures and spars for their size and for the period during which they were constructed.

It may be that this design contributed to the class having a reputation for excessive rolling. This reputation was borne out by Coast Guard Headquarters directing that the *Owasco*-class cutter *Sebago* have a "steadying sail" installed in late 1948 as an experiment to improve (i.e., reduce) the cutter's rolling. The experiment was a failure: There was "no noticeable improvement in roll characteristics" in moderate wind and sea conditions with the cutter making full speed and "no effect whatsoever on the amount of roll" while the cutter was hove to in fifty-knot winds on Ocean Station Charlie.

Sebago's captain, Cdr. Eugene A. Coffin, Jr., noted that it was not necessarily the idea of the sail but its implementation that was to blame. The sail size was too small to have any effect but too large for the cutter's rigging. After all, *Sebago* was not designed with the immense and heavy wire rigging required to support a spread of canvas. Captain Coffin pointed out that fittings that held the sail onto the standing rigging let go or wore significantly. Similarly, the running rigging (control lines) wore excessively, and blocks (i.e., pulleys with multiple sheaves) for the running rigging were smashed when the sail luffed.

Once the steadying sail idea was scrapped, Captain Coffin suggested that the Coast Guard install gyro stabilizers on the cutter. Captain Coffin was no fair-weather sailor who simply could not abide a little discomfort: He had twelve years of sea duty aboard a variety of Coast Guard cutters, all of it in the Atlantic and Arctic, and he said that *Sebago* had "a more irritating, nasty, lurching motion than any other vessel" he had been aboard. He reported his cutter as rolling as much as five degrees even while tied up at a pier, and when at sea,

"Her rolling is out of all proportion to the weather and sea conditions." His pleas fell on deaf ears. No gyro stabilizers were installed as the *Owasco*-class cutters rolled through the next two decades on ocean station.

The *Owasco*-class cutters also had a less-than-sterling reputation for reliability. The single-engine-room, single-propeller design no doubt contributed to this aura of fallibility because any single failure in the system meant the cutter would come to a stop. The records of these cutters are filled with reports of broken or burnt motors, generators, or shaft bearings and engine room fires that forced them into emergency repairs thereby taking them out of the ocean station operations schedule.

As with most legends, there is some basis in fact. The *Owasco*-class design did evolve during the course of the war. For instance, the seaplane that was originally to be carried amidships was abandoned. The Coast Guard did have plans to build a 316-foot cutter, which was to be a scaled-down version of the *Secretary*-class cutters, but it cannot be proved that the design was reduced in length even further to become the *Owasco* class.

There was also a reason for the number of *Owasco*-class cutters—thirteen. They were hull-for-hull replacements for three cutters that were grossly overdue for retirement (*Ossipee* and *Tallapoosa*, sister ships built in 1915, and *Unalaga*, built in 1912) and for the ten *Lake*-class cutters (2,075 tons, 250 feet) the service had turned over to the British as part of Lend-Lease in 1941.

Their lack of maneuverability (due to their single-screw configuration) mattered little while the cutters were on station and only served to heighten the cuttermen's pride in their shiphandling skills when they did get into tight-quarters situations.

Overall, the *Owasco*-class cutters were a mainstay of the Coast Guard's ocean station fleet, accomplished many rescues, and continued to sail on the ocean stations well into the 1970s.

* * *

TRANSIT PERIODS TO and from an ocean station were just routine passages for the cutters that conducted the patrols. They steamed alone, usually at a relatively high rate of speed. If heading toward station, a cutter would have squeezed out the last possible hour of liberty before sailing, leaving just enough time to arrive at station for an on-time relief. If heading home, everyone aboard was anxious to get there. Almost nothing short of emergency search and rescue would divert a cutter from its route. Such diversions were not completely unwelcome even for a homeward-bound cutter because they guaranteed a break in the routine.

The *Casco*-class cutter *Coos Bay* was heading home to Portland, Maine, from Ocean Station Bravo in February 1964 when just such a diversion disrupted her routine.

Almost 400 miles south of *Coos Bay*'s position, the British ship *Ambassador* had suffered a failure in her propulsion system. She was dead in the water. She had a significant list to port. Water appeared to be coming in through one (or more) of the cargo hatches. Her engine room was flooding. The wind was blowing a full gale. The seas were running in excess of thirty feet. These seas crushed the port lifeboat because the list put the boat close to the water. The starboard lifeboat was high and dry but useless because the port list kept it from being launched. *Ambassador* was in distress. Her radioman sent an emergency broadcast asking for assistance in the very early morning of February 18. Shortly thereafter, he sent an SOS.

The radio signals from *Ambassador* grew weak. The engine-room flooding meant the ship had no electrical power, and thus the radios were being operated on batteries that were not being recharged. It was only a matter of time before the radios, too, ceased to function. *Ambassador*'s crew estimated the ship only had eight hours before she would sink. Then, finally, the radioman told the world that he, along with the rest of the crew, was abandoning ship.

Coos Bay had already turned south, heading for *Ambassador.* The cutter tried to make top speed, 18 knots, but the course alteration put the seas on her beam and she began to roll heavily. Knowing that a capsized rescue ship would be of no use to anyone, *Coos Bay*'s captain, Cdr. Claude W. Bailey, reduced her speed to fifteen knots as she plunged and wallowed on.

AS DAYLIGHT GREW on February 18 and with one lifeboat destroyed and the other one useless, *Ambassador*'s crew of thirty-five took to their life rafts. Leaving a ship to enter a raft is an act of desperation.

Although a life raft can and is designed to survive heavy seas, it is also filled with peril. Because it has no means of propulsion, there is no way to control its direction other than holding it alongside the ship with line. And because it is made out of some type of pliable material, every minute the raft spends alongside a ship increases the probability the raft will snag on something and tear or that the fabric will be worn through, after which the raft will deflate.

Nor is it easy to get into a raft. The soft nature of an inflatable vessel gives poor footing. Once the raft is wet, which happens almost instantaneously after it is launched, the round sides nearly guarantee that a person is going to slip and fall, perhaps overboard rather than into the raft.

Unlike a ship or boat, a raft sits on the water rather than in it. Therefore, the raft will drift downwind at alarming speed. If it does not have any weight in it, the slightest breeze can pick up the windward side of the raft and capsize it. If the wind is a bit stronger, it can even cartwheel the raft downwind. If the raft does have weight in it, such as a ship's crew, stronger winds can still lift up one side and capsize the craft.

All of the foregoing is true in calm seas. With seas running to thirty feet, the problems increase dramatically. The waves will help the wind lift the side of the raft inducing capsize. The waves will break over the raft, filling it with water while guaranteeing no dry foot- or hand-holds for anyone. The waves can wash people in the raft out of the raft.

Yet, as long as it is undamaged, the raft will float, which is more than can be said for a ship that is listing and sinking. Therefore, *Ambassador*'s crew saw their choice as simple and stark: stay on the ship that they believed would go down or chance the risks of the rafts that would probably stay afloat.

The decision to abandon into the rafts was mitigated by the fact that an Italian cruise ship, the *Leonardo Da Vinci*, was on scene; *Ambassador*'s crew only had to stay in the rafts until the cruise ship could be brought alongside them.

Two large rafts and one small one were launched. The wind blew, the waves rolled, and the big rafts capsized, dumping thirty-two men in the water. The other three crewmembers, in the smaller raft, were blown downwind and out of sight along with the two fully inflated but now empty large rafts. Twenty-one members of *Ambassador's* crew swam and struggled and clawed their way back aboard their sinking ship. Eleven other crewmembers swam and struggled and clawed but did not get back aboard. The crew of *Leonardo Da Vinci* tossed lines over the side to pick the eleven men up, but exhaustion and the still crashing seas meant these men could not hold onto the lines to be pulled aboard. The sea had claimed the first of its victims from *Ambassador.*

Fearful of going inside the ship lest it capsize completely or sink suddenly, *Ambassador's* survivors huddled together on the bow and spent a long, dark night on the pitching, rolling derelict.

THE CREW OF *Coos Bay* spent a long night on the pitching, rolling decks of their cutter, still sailing at their best possible speed. They were already too late to save everyone but not too late to save someone.

A 10,000-ton Norwegian ore carrier, the *Fruen*, had joined the *Leonardo Da Vinci* at the scene. On the morning of February 19, in a notable display of seamanship, *Fruen's* captain approached *Ambassador* from downwind, held his own ship into the wind and seas and maneuvered close enough to *Ambassador* so that a line could be passed between the ships, and then held that position as some of *Ambassador's* crewmen were pulled through the water to *Fruen*.

Eventually, the movements of the two ships put enough repeated strain on the line so that it parted, but at least some of *Ambassador's* crewmen were safely aboard *Fruen*. *Fruen's* captain repeated his approach, made fast another line, and, two hours later, four more men made it across before this second line parted. *Fruen's* captain went back again, passed the line again, and started the transfer again. By the time this third line parted, a total of nine men had been pulled through the still-raging seas to safety, but twelve were still on the hulk of *Ambassador.*

Coos Bay had arrived in time to witness *Fruen's* remarkable feat. Because *Fruen* was succeeding, *Coos Bay* let the ore carrier continue her operations while the crew of the cutter directed other ships that were standing by to specific search areas in an effort to find the missing *Ambassador* crewmembers.

When *Fruen's* third line parted, she reported she had no more line to put out. It was time for *Coos Bay* to pluck the twelve remaining men off the foundering wreck of *Ambassador.*

The seas had not abated; in fact, the wind continued to howl from the northwest at forty knots and the seas, inevitably, stayed at twenty-five feet throughout the day as the wind was unabated in strength and unchanged in direction. The tops of these seas were being blown off by the same wind that pushed *Coos Bay* around.

The cutter was much smaller (2,500 tons) than either *Ambassador* (7,300 tons) or *Fruen* (10,000 tons). She also had a much shallower draft, which meant she would skid downwind far more quickly than would the other two ships. And she rolled consistently from twenty to thirty degrees and occasionally as far as forty-five degrees, the point at which it is easier to walk on the bulkheads than on the decks.

This extreme, even violent motion meant it would be impossible for *Coos Bay* to successfully launch and recover her boats. The boats would have been dashed against the side of the cutter or capsized or simply carried away by the wind and the seas.

Coos Bay could have followed *Fruen's* example and gotten close enough to pass a line to *Ambassador.* Although *Fruen* had been successful in getting nine men off the sinking ship by dragging them through the water, it had been a risky evolution: The men had frequently been submerged in the big seas and often disappeared from sight. Getting everyone or at least a large number of survivors into one raft seemed, if not safer, more likely to succeed.

Coos Bay nosed up to the bow of *Ambassador* and shot a line across. The line held, and *Ambassador's* crew pulled a fifteen-man raft to themselves.

THIS RAFT WAS designed, as are most large rafts, to keep people alive for days in the open ocean. Human beings on the open ocean have the same basic needs as those of human beings ashore: food, water, and shelter. These rafts, therefore, are equipped with rations and canned water. They also have a canopy, which is kept up by means of automatically inflated tubes. This canopy is how a raft provides shelter. If it is cold, being inside will provide some warmth as body heat is kept within the raft. If the sun is out, it provides protection from sunburn and heatstroke and helps reduce dehydration, the greatest threat to open-ocean survival.

The canopy, however, usually has one or, at the most, two flaps through which people can enter the raft. These openings are necessarily small to keep out the rain, wind, sunlight, and seas. Therefore, climbing through one of these openings is not easy under ideal conditions and is truly challenging under more difficult circumstances.

With the wind blowing a gale and the seas crashing all about them, the crew of *Ambassador* appeared reluctant to board the raft. The small craft pitched wildly on the seas and was difficult to control and keep alongside the ship. Five men finally jumped for the raft but could not, or did not, get inside the canopy. One among the waves that were sweeping the North Atlantic swept three men off the raft and into the sea. Other waves finished the job and now all five men were adrift.

Coos Bay broke away and headed for the man who had drifted farthest from the ship, the cutter's crew hoping that the others would either get back aboard the raft or *Ambassador*. Lookouts were directed to keep everyone in sight.

The crewmen executed a standard pickup for a man overboard, a drill they had practiced dozens of times. They rigged out an embarkation net ("scramble net") and had two swimmers, with lines attached, ready to go in the water. Then, Captain Bailey maneuvered his cutter upwind of the man, backed the engines so that she was making no way through the water, and thus slid *Coos Bay* to a stop alongside this first man. The swimmers went into the water, and with their help the man struggled onto the deck, exhausted but alive.

The lookouts reported one of *Ambassador*'s crew in the water very close to *Ambassador*'s stern. *Coos Bay* raced to him, the crew put over a line, which he was able to hold, and pulled him back to the cutter where the swimmers, once again, went into the water and helped him up to the deck. The other three men who had been on the raft and then in the water somehow made it back aboard *Ambassador*.

Coos Bay's attempt to recover most or all of *Ambassador*'s crew in a single effort using a raft was not a success: The cutter had recovered only two men. *Coos Bay* had watched *Fruen*, on the other hand, successfully recover nine men by simply pulling them through the water. It was getting dark, *Ambassador* was obviously on borrowed time, and leaving her remaining crew aboard the hulk one more night might well have killed them by exposure. *Coos Bay* decided to close right up on *Ambassador*'s bow, pass a line, and pull the remaining ten men to the cutter.

THE CUTTER GOT close, a line was passed, and a man fastened it to himself and jumped into the water from *Ambassador*'s bow. *Coos Bay*'s crew hauled away until the man was alongside the cutter, and once again the swimmers went into the water to help him aboard. Then, the crew did the same thing all over again and recovered another man.

One *Ambassador* crewman clings to the canopy of *Coos Bay*'s raft.

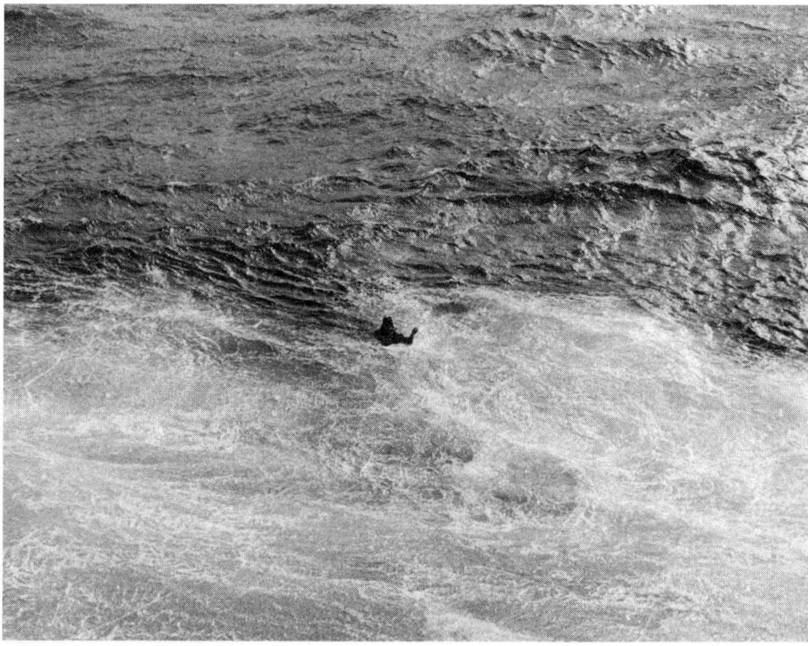

Coos Bay approaches an *Ambassador* crewman.

But *Coos Bay*, having much less draft and much more exposed superstructure than *Ambassador*, quickly drifted away from the stricken ship every time her crew carried out this evolution. The result was that in the time it took to haul a man alongside and get him aboard, the cutter had drifted so far away that a whole new approach had to be made to pass the line again. Time had joined the weather in becoming the rescuers' adversary.

Furthermore, *Ambassador*'s crewmembers were wearing their standard civilian life jackets, which did not hold the men's heads above water.[3] So the crewmen of *Coos Bay* decided that on their next approach, they would pass the line over with two Coast Guard life jackets that had collars; thus they could remove two men at a time and improve the chances that the men would make the transit without inhaling seawater.

Once again, *Coos Bay* struggled against the wind and waves, and once again her crew passed the line. This time the crew was able to bring two men aboard rather than one.

Each of these approaches and recoveries was a struggle in itself. The survivors were all nearly exhausted by the time they reached *Coos Bay*'s scramble net. They were barely able to grab onto the net, let alone pull themselves up to the cutter's main deck. Consequently, the Coast Guard swimmers found themselves struggling to help the survivors negotiate the treacherous last few vertical feet to safety.

During one of these recoveries, an *Ambassador* crewmember became ensnared in the net; his ladder to safety became a web of doom. On *Coos Bay*'s deck, Boatswain's Mate Third Class David P. Birchcrest, a standby swimmer, saw the man's plight and realized that his shipmates could not untangle the man. Birchcrest went over the

Coos Bay passes a line to *Ambassador*.

side without hesitation, a knife held in his mouth. He cut the man free, helped him to the deck, and went back to his standby duties. Birchcrest was later awarded the Coast Guard Medal for this instantaneous, intuitive, and heroic act.

COOS BAY WENT BACK again for a fourth pass and once again took off two men. All her crew had to do was make two more successful transfers and the job would be done, the lives of four more men saved.

Coos Bay got close again, and her crew passed the line again. Then, despite the fact that the four men remaining aboard *Ambassador* had watched this two-at-a-time technique work repeatedly, all four of them tied themselves into the line, rather than just two using the Coast Guard life jackets, and leaped off *Ambassador*. *Coos Bay*'s crewmen were stunned. But they had only one choice: haul all four men across the boiling seas.

As they did so, one of the men on the line appeared to lose consciousness and fall face first into the water. *Ambassador*'s four crewmen had tied themselves so far apart that none of the other three could get to this man and help him. *Coos Bay*'s crew hauled away and dragged this man aboard, but it was for naught. He never regained consciousness. He was *Ambassador*'s captain.

The other three were as good as can be expected under such circumstances. The cutter's crew gave them clothes, food, and places to rest. By the next morning, the eleven survivors were up and about the ship.

Coos Bay's job was not over, however. An aircraft had relocated the small raft into which three crewmembers had climbed—the raft had raced twenty-six miles away in just a few hours. By the time *Coos Bay* had completed her rescue operation and got to the raft, it was dark. A careful examination of the raft by searchlight confirmed everyone's fears: The raft was empty.

Throughout the night, *Coos Bay* remained on scene in a fruitless effort to find the missing men. By the next morning (February 20), the weather was deteriorating with the visibility dropping to four miles in rain and fog and the wind still blowing at forty knots. The probability of anyone remaining alive who had been in the water for more than twenty-four hours was near zero. *Coos Bay* was released from her search duties that evening and started pounding her way home. Two days later, *Ambassador* would finally sink.

AMBASSADOR WAS 7,300 tons, almost three times the size of *Coos Bay*. *Ambassador* sank whereas the cutter survived. *Ambassador* sank whereas *McCulloch* (*Coos Bay*'s sister ship) and *Owasco* (smaller still at 1,980 tons) survived conditions far more extreme than those that sank the larger ship.

Certainly, the designs of the ships played a role in these outcomes: The Coast Guard cutters had been subdivided and built to survive battle damage, whereas *Ambassador* had been built with large open spaces to maximize cargo space.

The cuttermen, however, had demonstrated their mastery of heavy-weather seamanship. And as the wind and waves literally tore *McCulloch* and *Owasco* apart, their crewmen not only survived but also were prepared to save the lives of others. In the end, it was not tonnage but tenacity, not size but skill that ensured survival.

ENDNOTES

1. This box analogy is imperfect and simplistic for describing the stresses on a ship's structure, but it suffices for the purposes of this book. Those interested in a more thorough explanation of hull dynamics should see any of the standard texts on naval architecture such as *Stability and Trim for the Ship's Officer* (LaDage and Van Gemert, 1983); *Modern Ships: Elements of Their Design, Construction, and Operations* (LaDage, 1965); or *Introduction to Steel Shipbuilding* (Baker, 1953).

2. After World War II, Coast Guard cutters patrolling ocean stations were under the operational control of the service's area commanders: Commander Eastern Area (Atlantic and Europe) and Commander Western Area (Pacific extending to Japan). In later years, COMEASTAREA became Commander Atlantic Area and COMWESTAREA became Commander Pacific Area. The administrative control of the cutters, however, belonged to the homeport Coast Guard Districts of individual cutters.

3. Most commercial life jackets have since been redesigned to keep an individual face-up even if unconscious.

4

CHARLIE

52°45' N, 35°30' W

IN 1947, AIRCRAFT STILL had names and not just tail numbers. The *Bermuda Sky Queen* (tail number NC18612), a Boeing 314 flying boat of American International Airways, departed Foynes, Ireland, for Gander, Newfoundland, Canada, at 1540 GMT on October 13 of that year. *Bermuda Sky Queen* would pass Ocean Station Charlie and obtain a radar fix from the *Secretary*-class cutter *Bibb* at 0205 GMT on October 14.

Later that morning, the young Captain Charles M. Martin and his flight crew realized they did not have enough fuel remaining to reach Gander, and their aircraft was past the point of no return. (Though strong headwinds were encountered during the flight, a U.S. investigation would conclude that at takeoff, *Bermuda Sky Queen* had insufficient fuel aboard to make the Foynes–Gander flight due to overloading, poor flight planning, and inaccurate power and fuel calculations, and the fuel problem was realized beyond the point of no return and not sooner because of poor flight monitoring.[1])

Four thousand feet below and a few hundred miles away, *Bibb* was holding her own against high winds and heavy seas on Ocean Station Charlie (then at 51°45' N, 35°30' W). It was difficult weather for a ship on the ocean, let alone a flying boat.

Captain Martin radioed Gander to explain his plight. He had aboard sixty-two passengers including children and seven crewmembers including himself for a total of sixty-nine. Gander thus radioed Ocean Station Charlie about the aircraft's status and the conclusion that the *Bermuda Sky Queen* would have to ditch.

At 0758 GMT, Captain Martin turned the flying boat around approximately 460 miles from Gander and 300 miles from *Bibb*. At 0914 GMT, the flying boat was about 100 miles from *Bibb*. *Bermuda Sky Queen* homed on the cutter's radiobeacon signal, and the crew of *Bibb* made ready for the ditching. At 0940 GMT, *Bermuda Sky Queen* began a series of four low passes over *Bibb*.

Airmanship and seamanship have a lot in common. There is a constant demand to watch the weather, a requirement to navigate prudently, an appreciation for spatial orientation, and a need to understand weight and stability. Aviators and sailors both work in a fluid environment, which means both are in constantly changing conditions and must adapt to whatever circumstances arise. It is not difficult, therefore, for sailors and aviators to understand each other's circumstances and to discuss those circumstances with a common language. Consequently, during the time between *Bermuda Sky Queen*'s decision to ditch and the landing itself, Captain Martin and the cutter's captain, Cdr. Paul B. Cronk, were able to plan how the ditching would occur.

The wind and the seas had not abated. Spray flew high into the air and over the wings and fuselage of the *Bermuda Sky Queen* as she touched down after the series of passes. Though the aircraft dipped one wing, it stayed together, and no one aboard was seriously injured in the landing. Captain Martin then maneuvered his flying boat toward *Bibb* to bring the aircraft alongside the cutter, secure a mooring line, and disembark the passengers and crew.

But things started to go wrong. The flying boat hit the cutter while *Bibb* was rolling thirty to thirty-five degrees in the rough weather, thus damaging the nose and wings of the aircraft. Putting the flying boat alongside the cutter would mean more damage to the aircraft, possibly breaking it up before people could get off. The question now was how to get the people off the flying boat and aboard *Bibb*.

Launching a small boat from the cutter in such conditions seemed problematic at best, disastrous at worst. Nevertheless, *Bibb* launched her port-side monomoy (i.e., pulling boat) to begin to recover the passengers and crew.

Meanwhile, the crew of the cutter spread *storm oil* as part of an effort to calm the seas. The practice of spreading oil was historically common in lifesaving procedures in heavy weather: The specific gravity of storm oil is less than that of water, thus the oil will float on the surface of the water, and as the surface tension of storm oil is less than that of water, the oil will suppress breaking wave crests, spray, and foam and thus smooth the surface of the seas.

Bibb's boat crew rowed over to the flying boat, observed its behavior on the sea, and made some practice approaches. After an hour, however, it became clear that the condition of the sea would cause the heavy boat to slam into the fuselage of the aircraft when alongside. This method of recovering the people from *Bermuda Sky Queen* would not work. *Bibb* brought her boat back aboard.

But not without incident. These boats were hoisted and lowered by two sets of falls. Each fall consisted of a block that could be attached to the boat (one forward, one aft) and another block mounted to a davit on deck. The lines running through these blocks would be taken in by means of a winch, thus raising the boat. Once the hoisting operation started, it had to proceed quickly to get the boat out of the water before it was smashed against the cutter's hull too many times.

As *Bibb*'s boat was hoisted, the forward falls disengaged. The boat was left dangling by the aft falls. Fortunately, there were lifelines that hung vertically from a span wire between the two davits. The boat's crewmen were able to hang onto these lifelines and get themselves back aboard the cutter. Eventually, the boat itself was wrestled back aboard.

Having eliminated the option of bringing the flying boat alongside and having decided that a boat recovery was not going to work, it seemed that the best and safest method of effecting rescue would be by pulling a raft through the water between the aircraft and the cutter. *Bibb* approached the flying boat, shot a line across, and attached the

Bermuda Sky Queen and *Bibb*'s pulling boat.

line to a raft so that the aircrew could pull the raft to them. As soon as the raft was clear of *Bibb*, however, the wind picked it up and tossed it through the air like debris. The line attached to the raft made it more like a kite than a craft, and *Bibb*'s crew realized that the raft would not work for this endeavor.

However, the weather was forecast to improve as the day went on. If they waited until later in the day, perhaps the moderating wind would allow the raft evolution to work.

THE FORECAST WAS WRONG. By 1500, the seas had only built, not diminished; 35-foot swells washed over the flying boat and the cutter, and the flying boat started to leak.

Darkness comes early at latitude 52 North in October. No longer could the evacuation be delayed. There was no guarantee that the flying boat would not sink in the coming night, and it would be foolish to allow the people in it to drown after they had survived the actual ditching.

In one of life's ironies, several of the passengers aboard the *Bermuda Sky Queen* were professional seamen: They were the captain and part of the crew of a U.S.-flag tanker, the *Chisholm Trail. Chisholm Trail's* captain, Thomas R. Quine, along with an able-bodied seaman named Gerald Harmon and a quartermaster named Arthur Brown volunteered to make an attempt: They would paddle one of the flying boat's rafts to *Bibb*.

Captain Cronk agreed, and when the men were ready, *Bibb* steamed upwind and laid down another oil slick in an effort to calm the still-raging seas. The men paddled literally for their lives but they could not reach the cutter. Once they were sufficiently clear of the flying boat, *Bibb* maneuvered close to them, threw them a line, and pulled them aboard. Everybody realized this worked once with three strong men in the raft but that it would never work for the less capable adults, let alone children, still aboard the aircraft. Everyone had been working at this problem all day, but only three of the *Bermuda Sky Queen's* sixty-nine people were aboard the cutter.

Bibb was running out of options. *Bermuda Sky Queen* could not come alongside. Floating a big raft from the cutter over to the flying boat had been a failure. Having passengers paddle small rafts across was hopeless. Attempting to use the cutter's boats was problematic. *Bibb's* boats were less than thirty feet long. Two of the boats had engines that took up considerable space inside the boat itself. The others were pulling boats that required several men at the oars. Neither type of boat had room enough to accommodate many passengers all at once. It would take at least seven individual round-trips between cutter and aircraft to get everyone off using a boat even if a boat could be launched and recovered successfully.

The crewmen knew that in the sea and wind conditions they faced, every time a small boat came alongside the cutter, it would be smashed against the hull of the cutter. No one knew how many such crushing blows these boats could take before they were literally pounded into pieces by the sea, but everyone knew that multiple approaches by any one boat was asking the impossible.

Furthermore, aircraft are relatively thin-skinned. As the cutter could damage the boat, the comparatively heavy boat could also damage the aircraft, possibly poking a huge hole in the fuselage and sending the flying boat and its people to the bottom.

So the cutter's crew decided to combine two techniques. They would launch one of the motorboats and tie a fifteen-man life raft to it. The crew of the motorboat would tow the raft near the flying boat, then float a line to the flying boat so that the raft, which was attached to the line, could be pulled to the flying boat; the passengers would board the raft; the raft with passengers aboard would be hauled back to the motorboat; the motorboat crew would take the passengers off the raft and onto the motorboat, leave the raft with the aircraft, take the motorboat back to the cutter, and repeat that process until everyone was off.

They launched the motorboat at 1636, one hour before sunset, and put a fifteen-man raft in the water alongside it. The raft broke free and drifted aft and into the cutter's propellers. This could have been truly catastrophic: The rubber raft with its various attached lines could easily have wedged itself into the cutter's shafts taking away its sole means of propulsion. *Bibb* would have been disabled and adrift in a raging North Atlantic while sixty-six people in a damaged aircraft were counting on the cutter to save their lives.

Undaunted, *Bibb*'s crew instantly put another raft in the water alongside the motorboat and then set about clearing the one from the stern. The motorboat departed for the aircraft with one raft in tow. Thirty minutes later, *Bibb*'s crew managed to pull the screw-threatening raft back aboard the cutter.

THE CAPACITY LABELS on life rafts are optimistic. A one-man raft is more oversized life jacket than craft. A ten-man raft will fit ten people, but there will be no secrets among them. Trying to board any size of raft in even the calmest of waters is a challenge. It would be difficult for fifteen people to get off a flying boat and fit themselves into a fifteen-man raft in 30- to 35-foot seas.

But in those conditions, a smaller number of people could more easily get into the raft at once, and they did. In three round-trips, *Bibb*'s small boat–raft combination managed to get twenty-eight of the *Bermuda Sky Queen*'s people from the flying boat to the cutter. Then the inevitable happened.

The repeated battering the motorboat took from being alongside *Bibb* had opened it up. The motorboat started to sink slowly as it made its fourth trip back to the aircraft. Furthermore, the raft was starting to deflate. It was approximately 1900, and it was dark. Sunset had occurred 1½ hours earlier, and the overcast guaranteed that neither moonlight nor starlight could penetrate the gloom. Although *Bibb* used her searchlights to provide some illumination, it was like turning on a flashlight in an unlit sports arena.

As the motorboat was heading back for the flying boat, the crew saw something unexpected, even unbelievable; the fifteen-man raft was adrift on the open ocean, well away from the aircraft. The painter, the line that held the raft to the flying boat, had finally given up and parted, setting the raft adrift. The motorboat crew had a sharp eye: It was dark, the raft was black, the wind was howling, and the seas obscured anything that was not close by.

And the raft had *sixteen* passengers in it. This not only exceeded the raft's capacity but also exceeded by five the number of passengers that *Bibb*'s crew had managed to get in the motorboat on the previous trips.

Nevertheless, the motorboat crew was not going to argue with success and thus did not depart from their previously successful methods. They started transferring passengers to the motorboat from the raft. Before they could finish, an unexpectedly big wave came out of the North Atlantic night and completely washed over the motorboat. The signal light was destroyed, the engine cover collapsed, the engine stopped, the gearbox cover was destroyed, and the propeller shaft fell out. The motorboat was not only without propulsion; it now had a hole, of size equal to the diameter of the shaft, below the waterline. The motorboat was already leaking from the damage incurred in its earlier beating against *Bibb*. Now, water was coming in from below with a vengeance, which meant that as the motorboat sank lower in the water, more and more waves would break over the side.

Bibb was about a mile away laying down more oil when Captain Cronk decided the motorboat had been gone just a little too long, and he headed his cutter over to investigate. Once he saw the motorboat, he knew instantly that he had to get his cutter alongside and alongside quickly. In a superior display of seamanship, he did it on his first approach.

While the motorboat crew had developed a system for transferring passengers from raft to motorboat, the rest of the cutter's crewmen had perfected their method for getting the survivors from the motorboat onto the cutter. A scramble net was lowered over the side of the cutter. Crewmembers would tie lines around themselves, work their way down the net, tie lines to the passengers, and help them climb up the net to the main deck. This system had worked on the three previous trips, and it was going to work again.

Except that in the three previous trips, the motorboat was not partially sunk and thus it did not have seas continually swamping it. This time, passengers and motorboat crew were washed in and out of the motorboat. No one could and no one cared to keep track of how many people went in the water, got back in the motorboat, and went in the water again.

The motorboat, or what was left of the motorboat, continued to be thrown against the side of the cutter. The motorboat's crew managed to pull one survivor from between the motorboat and the cutter just before he was crushed. Another passenger, a woman, was also swept from the motorboat into the void between it and the cutter. Boatswain's Mate First Class Ralph Keller, who was working on the net, jumped into the water and braced his back against the motorboat and his arms and legs against the cutter to keep the woman from being crushed. Keller then helped her into the relatively safety of the motorboat. Shortly thereafter, she along with everyone else ascended the net to safety.

Then both motorboat and raft sank.

THERE WERE STILL twenty-two souls aboard the battered *Bermuda Sky Queen*. *Bibb*'s motorboat was gone along with one of her rafts. Trying to float rafts between the cutter and the aircraft was impossible. Paddling a raft had not worked. It was well past dark. Neither the wind nor waves seemed to be abating. Yet a final, incredible effort to get everyone off the flying boat that night was made: *Bibb*'s crew launched the pulling boat from the pitching, rolling, heaving deck of the cutter in the middle of a gale in complete darkness.

It was a last-ditch effort and it did not work. The wind turned the boat crew's efforts into a folly. Each and every wave threatened the monomoy with swamping or capsizing. If the monomoy sank or rolled over now, in the darkness in the autumn North Atlantic, the men in it would die. They managed to get close enough to the flying boat to float a raft to it so that the people in the aircraft at least had a raft should the aircraft sink. But that was it. After an hour in 35-foot seas, the six men at the oars were exhausted. They struggled back to the cutter and were hoisted aboard.

The risks involved in a nighttime rescue are infinitely greater than those for a daytime rescue. All the elements of danger present in a daylight operation are amplified in the dark when men grope for things they cannot see and await the inundation of waves they know exist but that seem to creep up out of nowhere. With most of the *Bermuda Sky Queen*'s people out of the flying boat, the aircraft seemed to be somewhat more buoyant and less likely to sink. Admittedly, the ride was neither pleasant nor safe, but the flying boat seemed to be in no immediate danger. The fact that it had remained afloat since ditching in the morning gave credence to the possibility that it might survive the night.

Everyone in the flying boat and all those on the cutter had spent the daylight and nighttime hours watching the repeated near-catastrophes involved in the rescue effort: the rafts that flew around on the wind, the agony of the three men who tried to paddle a raft to safety, the destruction and sinking of the motorboat, the struggles with the pulling boat. With this as well as the dangers involved in a night rescue in mind, the captain of the *Bermuda Sky Queen* and the captain of *Bibb* decided to wait until morning to finish the evacuation. Of course, if the flying boat started to break up during the night, the rescuers would take to the oars, but at least the aircraft's people had a raft, and prospects looked best for a morning rescue.

The storm raged throughout the night while twenty-two people suffered the North Atlantic's worst in the hollow fuselage of a flying boat. When dawn finally broke on October 15, the previous day's forecast for moderating weather had come to pass. Captain Cronk decided to launch his one remaining motorboat, a captain's gig. The crew of the gig made one successful run and rescued eight passengers, but the gig suffered mechanical failure on the second run. It was back to the tried and tiring pulling boat. The crew of the monomoy now had success and would rescue the remaining fourteen individuals from the flying boat. Meanwhile, the gig was restarted by its crew and brought back aboard *Bibb*.

Everyone was saved. It had taken just about every conceivable method of rescue to create this success: a raft floated across on a line, another raft paddled by men, motorboats towing rafts, pulling boats towing rafts. But it worked; everyone was alive, and no one was seriously injured.

Captain Cronk and Captain Martin now considered attempting to take the flying boat under tow: it had survived a tremendous beating but was still afloat. The weather had moderated, and they thought that the cutter could now get alongside the aircraft. Their better judgment prevailed, however, and they realized it would be impossible to tow the broken aircraft a thousand miles to safety. Instead, with the concurrence of Captain Martin, *Bibb*'s crewmen used their weapons to sink the *Bermuda Sky Queen*.

Then *Bibb* headed for Boston to offload her unexpected passengers. A few months later, the nation's government showed its gratitude. Petty Officer Keller, who had leaped into the water to save the woman from being crushed, was awarded the Gold Lifesaving Medal, and Lt. (jg) Clarence S. Hall, who had been the boat officer during the four successful round-trips evacuating forty-four passengers (including the ill-fated and extremely dangerous final trip by motorboat), also was awarded the Gold Lifesaving Medal. Five other boat crewmen, Lt. (jg) Bernard Brown, Machinist Lindel Hall, Gunner's Mate First Class John Johnston, Chief Motor Machinist Harry LaFever, and Machinist's Mate Third Class Philips Taylor, each were awarded the Silver Lifesaving Medal.

Before that, however, *Bibb* had given herself a more humble but no less significant award for her arrival in Boston: a broom affixed to the top of her mast, an old sailor's symbol for a "clean sweep," here telling the world that the cutter's crew had managed to save every person aboard an aircraft that might have otherwise carried all to a watery grave.

BIBB HAD ANOTHER clean sweep less than a year later.

In mid-September 1948, a hurricane packing 100-knot winds spun along the northern part of the Gulf Stream about 300 miles south-southeast of Newfoundland. As it moved northeast, the storm's swells lashed Newfoundland. Between the eye of the storm and the Canadian coast was the cutter *Bibb*, bound from Boston for a fuel stop in Argentia, Newfoundland, on her way to Ocean Station Bravo.

On the other side of the storm, the Portuguese fishing schooner *Gaspar* reported she was sinking shortly after 1300 on September 15. *Bibb*'s commanding officer, Capt. Donald G. Jacobs, did not hesitate; at 1310, he altered his cutter's course toward the old schooner.

All night, *Bibb* struggled through the wind and seas. Fortunately for both the cutter and the fishing schooner, the hurricane was moving quickly northeast at thirty knots, which meant conditions continued to improve as night turned to day. Finally, with the help of a Coast Guard aircraft, *Bibb* found *Gaspar* shortly after 0915 on September 16.

At 1147, *Gaspar*'s captain decided to abandon ship. Despite the seas, which were still in excess of twenty feet, the crew of *Bibb* put two boats in the water, each with a raft preparing to take at least some of the crew off the schooner. *Bibb*'s small boats towed the rafts alongside *Gaspar* getting thirteen men in one raft and fifteen in the other. Within forty minutes, both rafts were back alongside *Bibb* successfully unloading the stricken fishermen. One of the boats struggled back to *Gaspar* and picked up the remaining twelve crewmen. By 1328, all forty of *Gaspar*'s crew were safe aboard *Bibb.*

But the schooner did not sink. It was in sorry shape and undoubtedly taking on water, but it continued to drift around making itself a hazard to other vessels in the area. With the concurrence of the schooner's captain and the Coast Guard's Commander Eastern Area, the crew of *Bibb* shot at the schooner until it finally gave up, rolled over, and sank on the morning of September 18. *Bibb* then steamed toward Newfoundland, and the forty fishermen were delivered to the Portuguese consul in Argentia the next morning.

Bibb and her crew had saved the lives of 109 people on the high seas in eleven months while losing none of them. It was a singularly impressive record that would not be equaled by any of the other ocean station cutters.

Bibb would take on fuel in Argentia on September 19 and then resume her voyage for Bravo.

* * *

BIBB AND HER SISTER SHIP *Duane* were the first two cutters assigned to the weather observation stations when the Atlantic Weather Observation Service began in February 1940. In 1977, yet another cutter of the same class, *Taney*, would be the last Coast Guard cutter to patrol an ocean station. The *Secretary*-class cutters and their crews would write a lot of the Coast Guard's history for the period 1940–1977.

In 1935, the Coast Guard began construction of these major cutters. For years, the service had operated mostly (as it does today) with vessels less than 100 feet in length. This is a consequence of the Coast Guard having been created from the merger of the U.S. Life-Saving Service (itself an amalgamation of various state and local lifesaving services) with the U.S. Revenue Cutter Service in 1915.

The Life-Saving Service consisted of small-boat stations that were located along all the coasts of the United States including those of the Great Lakes. These stations had at least one and usually two or three boats designed to be launched from shore by the surfmen to rescue crews from stranded and/or sinking ships. The Revenue Cutter Service, on the other hand, was made up of cutters that worked offshore, enforcing customs laws and also performing rescue work. The Revenue Cutter Service generally had fewer than forty-five cutters, a number that was greatly counterbalanced by the more than 140 Life-Saving Service stations and their surfboats.

The Revenue Cutter Service and its successor Coast Guard had a well-earned reputation for using ships that were old and ships that were cast-offs from the Navy. For instance, the venerable cutter *Bear*, launched as a sealer in 1874 and commissioned as a cutter in 1885, was on active duty with the service until being decommissioned in 1929, after which she served Adm. Richard E. Byrd, USN, in two Antarctic expeditions. After her return from the Antarctic in 1941, *Bear* was assigned to the Greenland Patrol under Navy commissioning, and in 1944 she was removed from active service.

In 1919 and 1920, the Coast Guard took delivery of almost three-dozen former Navy vessels, a few of which remained in the Coast Guard's fleet until the late 1930s. In 1935, the year the first *Secretary*-class keels were laid down, there were at least six former cutters of the Revenue Cutter Service (i.e., cutters that were more than twenty years old) in the Coast Guard's fleet.[2]

It was refreshing and promising, then, for the service to begin acquisition of a unique, new fleet during the mid-1930s. The service had already acquired ten new *Lake*-class cutters, which were built and commissioned between 1928 and 1932, but the *Secretary*-class cutters had even greater promise.

These cutters would be modern, fast (for the time), well constructed, and durable. They were so durable that, of the seven built, six stayed on active service with the Coast Guard until being individually decommissioned across the period 1980–1988, *Ingham* and *Taney* being the last of these six to be decommissioned.

Displacing 2,350 tons over their 327-foot overall length, the design of these cutters was based on the Navy's *Erie*-class gunboats. They were propelled by high-speed geared steam turbines that provided each of two shafts with 3,100 horsepower for a top speed of more than twenty knots. This high speed and maneuverability (due to the twin-shaft configuration) would make these cutters valuable as submarine chasers and convoy escorts during World War II.

The Coast Guard was part of the Department of the Treasury in the 1930s. This was a result of the Revenue Cutter Service having been part of the U.S. Customs Service within the Department of the Treasury ever since the creation of the Revenue Cutter Service in 1790 and of the Life-Saving Service having been in the same department since its creation in 1878. Consequently, these new cutters were to be named for former secretaries of the Treasury and would be known as *Secretary*-class cutters.[3] The first cutter, *Alexander Hamilton*, was named for the first Treasury secretary—the man who had been instrumental in founding the Revenue Cutter Service. The other cutters of the class were named *George M. Bibb*, *George W. Campbell*, *William J. Duane*, *Samuel D. Ingham*, *John C. Spencer*, and *Roger B. Taney*.

Earlier cutters had also been named for Treasury secretaries. There had been at least three earlier *Bibb*s, four *Campbell*s, two *Duane*s, three *Hamilton*s, three *Ingham*s, one *Spencer*, and one *Taney*. This re-use of names would continue for the Coast Guard into the late 1980s when new, *Famous*-class cutters (1,800 tons, 270 feet) were launched with the names *Campbell* and *Spencer*.

Confusing matters even more, in mid-1937 all the *Secretary*-class cutters had their names shortened to last names only (*Bibb*, *Campbell*, etc.), though the cutters would frequently be referred to using their original full names as late as 1947.[4] For instance, in a 1947 letter of appreciation, the passengers of the *Bermuda Sky Queen* declared "that the behaviour and skill of the Captain and crew of the USCG cutter *George M. Bibb* . . . are worthy of the greatest commendation."

Finally, these cutters (along with the other ocean station cutters) were inevitably and almost universally referred to within the Coast Guard by their length; that is, as "327s."

THE WAR-FIGHTING capabilities of the *Secretary* class would make the cutters too useful as convoy escorts for them to continue to be stationed as weather ships on Stations 1 and 2 during World War II.

Secretary class: The cutter *Ingham.*

Duane on North Atlantic convoy duty.

This escort duty would lead to the only loss of one of these cutters. Convoy escorts not only fought U-boats but sometimes had to aid stricken vessels as well. On January 29, 1942, *Hamilton* had been towing a disabled Navy auxiliary ship and had just released the tow to another vessel about 10 miles off the coast of Iceland when an explosion ripped through her engine room killing everyone in the space and leaving a gaping hole in her side. A torpedo fired by *U-132*[5] had delivered the fatal blow. *Hamilton* began to settle and list, and the crew started a hurried but orderly abandonment.

Two attempts were made to take *Hamilton* in tow, as it seemed she could be saved. However, she rolled onto her beam ends then capsized while under the second tow and finally slipped beneath the waves during the night of January 30–31.

Hamilton's sisters would avenge her death.

While escorting convoy ONS-152 off Iceland, *Ingham* attacked a sonar contact on December 17, 1942. Although her crew could not confirm a kill at the time, Kriegsmarine records indicated that *U-626*[6] was sunk near the site of *Ingham*'s attack, and she was later credited with the kill.

On the evening of February 22, 1943, while escorting convoy ON-166 in the North Atlantic, *Campbell*'s radar picked up *U-606*[7] on the cutter's starboard bow. The U-boat had earlier been subjected to a depth-charge attack by other convoy escorts. *Campbell* headed straight for *U-606*, bringing it under fire from deck guns and depth charges. *Campbell* got closer than she intended, however, and a collision with the U-boat's bow plane sliced the cutter's engine room open, flooding it, and causing *Campbell* to lose all power. *Campbell* was subsequently towed to safety, repaired, and returned to the fight in the North Atlantic. *U-606* had been sunk as a result of the combat with *Campbell.*

While also escorting convoy ON-166, *Spencer* is believed to have sunk *U-529* a day earlier on the evening of February 21.[8]

On April 17, 1943, while escorting convoy HX-233 in the North Atlantic, *Spencer* began a depth-charge attack against *U-175*.[9] *Duane* arrived to assist, and the U-boat was soon brought to the surface where it lost a deck-gun battle with *Spencer.*

Later in the war, the *Secretary*-class cutters would be converted to flagships for amphibious operations with an increase in communications equipment and ever-changing weapons configurations. They would continue to serve in combat zones until the last days of the war.

Their impressive wartime record gave these cutters a well-deserved reputation for great sea-keeping ability. When the crew of *Bibb* made their spectacular rescue of the *Bermuda Sky Queen*'s passengers and crew in 1947, the reputation of the cutters as strong, capable, seaworthy ships was sealed. It would be another thirty-nine years before the last of the *Secretary*-class cutters made its final voyage under the Coast Guard ensign.

* * *

SIX WEEKS BEFORE *Bibb*'s rescue of the *Bermuda Sky Queen*'s passengers and crew, another dramatic rescue had occurred on Ocean Station Charlie.

On September 1, 1947, the crew of the *Casco*-class cutter *McCulloch* received a radio message warning of a whole gale due to pass over Charlie and thus *McCulloch*. The cutter was already experiencing increasing winds, and once the message was received, *McCulloch*'s captain, Cdr. W.C. Hogan, directed his crew to make preparations for heavy weather: ensuring everything was tied down, lashed up, properly secured, and lifelines rigged on deck. Captain Hogan ordered his cutter headed into the seas to ride out the storm. As was always the case in these conditions, *McCulloch* pitched as she made way and rolled consistently as much as thirty degrees.

By that evening, the wind was blowing fifty knots and more, and the seas had built to thirty feet. As night fell, the conditions continued to deteriorate with heavy rain and driving spray cutting visibility to near zero. One of the crewmembers said to no one in particular, "Anyone falling overboard in that sea is just a lost man." No one argued with him.

Weather conditions, no matter how severe, never stopped the requirement of the cutters to keep taking, analyzing, and transmitting weather information. All the logbooks of the ocean station cutters show them launching and tracking weather balloons and taking other meteorologic observations even when they were engaged in significant search-and-rescue cases. The weather balloons were launched every six hours, and the cutter crews recorded the weather hourly. But the U.S. Weather Bureau observers aboard the cutters made more detailed observations, and they did so around the clock. Thus, even at 0200, it was likely that one of the weathermen would be working in his space, completing observations, filing reports, or checking on the operation of his equipment.

Aboard *McCulloch* at 0200 on September 2, the senior weather observer, Mr. R.W. Mosher, found one of his instruments malfunctioning. As was the standard practice, he asked for one of the cutter's electronic technician's mates to help him fix the device. Electronic Technician's Mate Third Class Robert T. Gray went aft to the weather shack to look at the instrument. He decided he needed some tools and a manual if he was going to be able to make repairs. To get those things, he had to go forward to the radio room.

That meant he had to go outside. Rain was coming down in sheets. The cutter continued to pitch and roll. Gray went out the watertight door of the weather shack, battled his way forward, got back inside the cutter's superstructure, retrieved his gear from the radio room, and started the return trip. Once again, he threaded his away along the rain-soaked, pitching, heaving, rolling deck aft to the weather shack. He opened the port-side watertight door into the shack. He was almost safe once more inside the cutter.

Mosher saw the door open and saw Gray's hand holding onto the door. Then *McCulloch* took a hard roll to starboard. The heavy door swung closed as Gray struggled to control it. Then *McCulloch* took a hard roll to port. Now the door, which an instant before had been Gray's entrance to safety, became his nemesis. It swung open knocking Gray off his feet. The rain-slicked deck gave him no chance to stand up or regain any kind of hold on his cutter. As *McCulloch* continued her roll to port, Gray slid under the lowermost lifeline and tumbled into the North Atlantic. In the dark. In the impenetrable rain. In fifty knots of wind. In thirty-foot seas.

Mosher jumped up, stepped outside, and looked around for Gray. He saw nothing, but he heard a voice holler out of the stormy night. He told another crewmember to throw a life ring over the side as he called the bridge to report a man overboard.

The watchstanders on the bridge did not hesitate. One individual announced "Man Overboard" over the cutter's loudspeaker system as others threw life rings over the side. Captain Hogan raced to the bridge and immediately relieved the watch officer so that he, the captain, could take direct control of the recovery actions.

THE LIFE RINGS that *McCulloch*'s crewmen desperately threw to their shipmate had waterproof lights attached to them. In this type of life ring, one light is spliced to the life ring by a short piece of line. Onboard, the life ring hangs in a simple bracket, and the light is held, inverted, in a bracket next to the life-ring bracket. The light is inverted because it has a simple, weighted switch that keeps the light off while it is upside down but that turns the light on when it is right-side up. The battery is in the bottom of the light, and its weight ensures that the light will activate: The light will float upright in the water.

The lights are, therefore, mechanically activated electrical devices that get tossed into a saltwater environment. Sometimes the switch fails. Sometimes water gets past the watertight seals and shorts the circuit. Sometimes the casing of the light has a minute crack that allows water in and shorts the circuit. Sometimes the spring-loaded battery contacts fail. Sometimes big seas smash them.

No one kept track of how many life rings were thrown overboard. It was at least three, and probably several more. Of them all, only one had a light that functionally survived the trip over the side and into the frothing waters.

Gray saw that light, his beacon of hope, and he swam for it. As he grabbed the life ring, the light went out. At least he had the life ring so he could stay afloat as long as he had strength to hold onto it. No one on the cutter, however, could see him, and now no one on the cutter could see the single light, which would have given them something to aim for in trying to pick him up.

Crewmembers had turned on searchlights and swung their beams toward the life-ring light. Before the searchlights could be properly aimed, however, the life-ring light went out. Now the searchlights swept about the sea, their luminosity reduced by the rain and spray as they searched for the missing man.

The next few minutes were filled with decision-making and actions. One wrong decision, one improper or incomplete action could have resulted in Gray's death. *McCulloch*'s crew could afford no errors.

And they made none. Captain Hogan had already decided that launching a boat to recover Gray was out of the question. The ship was still rolling and pitching heavily, and there was every reason to believe that trying to put a boat in the water under those conditions in the dark and rain would, at best, severely damage or destroy the boat or, at worst, dump the entire boat crew into the water as well. If that happened, *McCulloch* would then have six men overboard instead of just one.

HAVING DECIDED TO pick up Gray by using the cutter rather than the boat, Captain Hogan now had to decide, in just a few seconds, which method he would use to get alongside the man.

McCulloch was, of course, under way, but more importantly she was making way, that is to say she was moving ahead through the water. Standard seamanship texts say that when a ship is making way and has a man overboard, the ship should execute a turn to get alongside the man. The texts say this because the presumption is that once the person is in the water, he or she stops while the ship continues to move ahead. Thus the person in the water is soon astern of the ship and so the ship must turn around and go back to where the person is.

Standard seamanship texts, however, were not and are not written for fifty-knot winds and thirty-foot seas. Because *McCulloch* was struggling to keep her bow into the seas and thus was sailing into the wind, she was barely making way through the water even though the shafts were turning for about five knots. Under those conditions, Gray was not falling rapidly astern, though the cutter was moving slowly away from him.

Furthermore Captain Hogan, along with everyone else, had watched the life rings go over the side and knew that only one life-ring light was working. If that light failed during the few minutes it would take to complete a turn, McCulloch might never relocate Gray again. There was simply no time to "waste" executing a turn.

Therefore, Captain Hogan chose to stop his cutter as quickly as possible. Simultaneously, he intended to turn or twist the cutter to port where it was so that *McCulloch* would stay close to Gray and also stay upwind of him. Physics dictates that any stopped ship in any amount of wind will drift faster than any person in the water: the ship's side exposes greater surface area (known as *sail area* to seamen) to the wind than does the head of a human being. Thus there is greater pressure on the ship and it will drift faster than the person. If the ship is stopped upwind of the person, the ship will, inevitably, drift down on him. That was Captain Hogan's plan.

Meanwhile, on deck everyone was turning out and making preparations to pick up Gray whenever the cutter got alongside him. *McCulloch*'s executive officer, Lt. Cdr. Clifford S. Gerde, the second in command, was on the main deck directing crewmembers to get into rubber swim suits, an early version of today's wet suit.

The idea of going into that ocean, even with a "proper" suit on, had to give pause to everyone on deck. *McCulloch* was rolling and pitching hard, the wind was howling, violent rain obliterated everything, sheets of spray were breaking all over the cutter, waves themselves were breaking across the aft deck, and the hazy illumination of the searchlights could not break the night.

Without a moment's hesitation, Boatswain's Mate First Class Marion K. Reynolds told Lieutenant Gerde he would put on a suit, and, assisted by a couple of his shipmates, he did. Reynolds had a tending line attached to him, and then he and the other swimmers stationed themselves along the deck as everyone kept searching the sea for Gray.

ON THE BRIDGE, Captain Hogan got his cutter turned by using maximum engine power ahead on the starboard shaft and astern on the port shaft. Once the life-ring light extinguished itself, he had no method of figuring out where Gray was other than his own sense of the sea, but he believed that he had gotten *McCulloch* almost directly upwind of the missing man. He took all the way off his cutter and hoped that as she drifted down, she would drift onto her missing man.

That was the good news. The bad news was that *McCulloch* was now beam to the wind and seas, which meant the huge waves crashed against the cutter's starboard side inundating the decks and rolling the cutter severely. It was possible she could roll far enough to port that the seas on that side would find their way into vents or through partially open doors and begin to sink the cutter. It was possible she could take a lurching roll and spill another man or men into the ocean. It was possible she could roll all the way over and capsize. To save one, all had to be put at risk.

Someone heard Gray yell. Then everyone heard him yell. The searchlights were redirected and after a long minute finally illuminated him. He was swimming, feebly, toward the cutter, which was about seventy-five yards away from him. As the distance closed, Gray yelled out again, this time saying he could not last much longer.

Reynolds immediately asked Lieutenant Gerde for permission to enter the water. Reynolds had no guarantee that he could get over the side without being washed away by the breaking seas. Reynolds had no guarantee that he would not be knocked unconscious or have his arms or legs or back broken as the cutter rolled hard while he tried to get over the side. Reynolds had no guarantee that he could reach Gray. Even if he did reach Gray, Reynolds had no guarantee that he could get both of them back to the cutter. Even if he did get them both back to the cutter, there was no guarantee they could get back aboard without breaking bones or suffering concussions.

Reynolds asked for no guarantees. The other swimmers might have been willing to take the same risks, but Reynolds stepped forward first and without hesitation to save his shipmate. Lieutenant Gerde granted permission.

Gray was almost completely exhausted and incapable of helping himself by the time Reynolds reached him. Reynolds wrapped his arms around Gray and kept him afloat as other crewmembers dragged the two back to *McCulloch*. Then, as the crew hoisted the two of them up the slick, rolling, hard steel side of the cutter, Reynolds used his own body to keep Gray from getting battered. Another boatswain's mate, Boatswain's Mate Second Class William P.F. Marter, lowered himself partway over the side to help Gray and Reynolds get back aboard. In a matter of a few moments, they were all aboard, inside, and alive. Gray and Reynolds were taken to *McCulloch*'s sick bay where they were treated for shock and exposure.

When he filed his official report, Captain Hogan commended his entire crew for the "notable absence of confusion and a general appreciation of the problems involved and intelligence displayed in their solution." He further commended several individual members of the crew including BM2 William Marter for the personal risk he undertook in getting Gray and Reynolds back aboard and Lieutenant Gerde for his "intelligent handling of the lifeline," which got both men aboard without further injury.

Captain Hogan recommended Reynolds for the Gold Lifesaving Medal. He had earned it. He deserved it. He received it on March 2, 1948.

GRAY'S RESCUE and the rescue of the passengers and crew from the *Bermuda Sky Queen* had been completely successful: all lives saved and no one seriously injured in the process. Sometimes, however, rescue efforts are less than completely successful.

The *Casco*-class cutter *Chincoteague* was on Ocean Station Charlie on January 10, 1957. The weather was typical for this station at this time of year: overcast, rain, 20- to 25-knot winds, and seas building to 15 feet.

Shortly before noon, the Finnish collier *Finntrader*, loaded with coal, was crossing the northern extremities of Charlie when she suffered an explosion in her engine room. The first, second, and third engineers along with one of the unlicensed engine-room crewmen suffered serious burns. *Finntrader* needed immediate medical help and sent a radio message calling for it.

Within three minutes of receipt of *Finntrader*'s call for help, *Chincoteague*'s captain, Cdr. Raymond G. Miller, put her on a course for the stricken ship. As the cutter struggled through the sloppy seas, the crew was able to offer medical advice to those aboard *Finntrader*. This was because *Chincoteague* had a doctor aboard, a uniformed member of the U.S. Public Health Service. Ocean station cutters almost always had a doctor aboard. The doctors, as was true with the weathermen, were assigned to a cutter only for the duration of the patrol; they were not full-time members of the cutter's crew. They were, however, physicians and thus able to provide more extensive medical care than could the cutter's corpsman, who was a full-time crewmember.

Chincoteague struggled through the seas all night, finally rendezvousing with *Finntrader* on the morning of January 11. All of the burn victims were in poor condition and thus confined to stretchers. It appeared that the only way to get the injured men off *Finntrader* and aboard *Chincoteague* was by using the cutter's motorboat.

Any launch and recovery of a small boat is a challenge. As weather conditions get worse, the challenge becomes greater because the rising seas will slam the small boat against the ship, potentially damaging the boat and knocking the crew into the water.

Furthermore, the motorboats aboard cutters in the 1950s were designed and built with the engines under covers (really just large boxes) located in the middle of the boat. Although this was fine for boat crews, who could sit all around the engine box, it left very little open space in the motorboat. This meant that carrying a stretcher, and especially a person in a stretcher, would be awkward at best. Therefore, *Chincoteague*'s crew would only be able to carry one victim at a time.

In other words, four round-trips between the cutter and the collier would be required. *Chincoteague*'s small boat would have to make eight different approaches alternating between their own cutter and *Finntrader*. That was eight opportunities to have the motorboat battered, the crew pitched overboard, or the stretcher-bound victims further injured.

Chincoteague launched her motorboat with a six-man crew: Ens. Peter D. Corson, Engineman First Class David Steele, Quartermaster Third Class William G. Hellweg, Boatswain's Mate Third Class John J. Parra, Boiler Tender Third Class Larry A. Barker, and Seaman Walter T. Hasenzahl. In addition, the doctor, Lt. Cdr. Barnard L. Kaye, and *Chincoteague*'s own corpsman, Chief Carl F. Baker, also went into the boat to begin the shuttle, transfer, and care of the grievously burned *Finntrader* crewmen.

They did it. Less than two hours after arriving on scene, *Chincoteague* had all four victims aboard and headed for the hospital in St. John's, Newfoundland. That was the good news. The bad news was that *Finntrader*'s second engineer had been so badly burned that he was dead by the time he was brought aboard *Chincoteague*. The unlicensed crewman was burned almost as badly, and Doctor Kaye held out little hope for his survival.

Chincoteague needed to make a transit of more than 1,000 miles to reach St. John's. The wind had built to fifty knots out of the southeast by the early afternoon of January 11, which meant *Chincoteague* had the ever-increasing seas on her beam as she steamed on her southwesterly course. Thus she rolled badly. By the next morning, the wind had shifted to the southwest, reducing the roll but making the cutter pitch heavily into the oncoming seas. Nevertheless, Doctor Kaye and Chief Baker along with several helpful shipmates labored away at keeping the three remaining men alive.

This was a real challenge. Victims, whether they are burn victims or are suffering from other injuries, are usually brought quickly to someplace where they will not have to be moved. But these three men were inside a cutter that was first rolling then pounding its way across the winter North Atlantic Ocean. This made it more painful for the patients and more difficult for the doctor and corpsman, who had to carry out duties requiring finesse and gentleness without knowing exactly when the cutter would take a hard roll or pitch awkwardly.

They lost the fight for the life of the most severely burned man as *Chincoteague* crashed through the waves headed for Newfoundland. *Chincoteague* pressed on, making her best possible speed but often having to slow down because of the raging seas.

Finally, on January 17, the cutter arrived at St. John's where she offloaded the two surviving men as well as the two deceased. It was not a completely successful rescue, though it was not for a lack of effort. The hazardous boat trips, the difficult medical care given under poor conditions, and the six days of rolling and pounding across hostile seas would have challenged even the best of seamen.

The Lifeboat Society of Finland understood all those challenges and decided to recognize the efforts of *Chincoteague*'s crew by presenting a bronze plaque to the cutter. In October, Finland's ambassador to the United States, Richard R. Seppala, made the presentation to Commander Miller at the cutter's homeport in Norfolk, Virginia.

The ambassador noted that efforts such as that of *Chincoteague's* crew were not well known outside the maritime community—yet there can be no higher praise for lifesavers than that given by other lifesavers.

ENDNOTES

1. Civil Aeronautics Board (1948). *Bermuda Sky Queen* had been loaded more than 5,000 pounds in excess of its maximum permissible gross weight of 84,000 pounds.

2. Little changed in succeeding decades. The *Casco*-class ships were all former Navy vessels that saw service as Coast Guard cutters on ocean stations for almost thirty years. The *Hamilton*-class cutters, which were built for the Coast Guard between 1965 and 1972, were all still in service thirty-seven to forty-four years after their construction.

3. Various Coast Guard sources list these cutters as *Treasury* class, *Hamilton* class, *Secretary* class, or 327 class. For no reason other than consistency, they will be referred to throughout this book as *Secretary* class.

4. *Hamilton* would also have its name changed back to *Alexander Hamilton* when the Coast Guard was put into the Navy Department during World War II to avoid confusion with a Navy vessel bearing the same name.

5. Rohwer (2005, p. 140).

6. Rohwer (2005, p. 219).

7. Rohwer (2005, p. 232).

8. After World War II, *Spencer* was credited by the U.S. Navy for the sinking of *U-225* on February 21, 1943. However, the British have credited the sinking of *U-225* to Liberator "S" of RAF No. 120 Squadron on February 15, 1943. (Source: "U-225." U.S. Coast Guard Historian's Office, Washington, D.C.) Rohwer (2005, p. 232) states that *Spencer* is believed to have sunk *U-529* on February 21, 1943.

9. Rohwer (2005, p. 246).

5

DELTA

44°00' N, 41°00' W

FOR CENTURIES MEN HAVE gone to sea in ships, and for centuries they have met with disaster. The same type of weather that battered the Coast Guard cutters on ocean stations also wreaked havoc with commercial ships on routine transits near those stations.

The Coast Guard's long history as a lifesaving organization gave confidence to mariners needing assistance. The cutters were manned by personnel from a variety of backgrounds within the service, and more than a few of each crew had earlier experience in rescue work. That experience and talent, the standardized search-and-rescue procedures that the cutter crews had to understand and the training that they were required to complete, and the deployment of cutters far offshore meant that mariners crossing the ocean had at least some assurance that help was not too far away when and if catastrophe struck.

For the German collier *Helga Bolten*, such help was about 130 miles away on the night of October 29–30, 1956. *Helga Bolten* was slightly west of the western edge of Ocean Station Delta that stormy night, bound for Bergen, Norway, from Norfolk, Virginia. She had been battered by a storm for more than twenty-four hours, and as the clock crept past midnight, winds of forty to forty-five knots had continued to pile up seas that were at least twenty feet high and that occasionally peaked at forty feet.

Because *Helga Bolten* had a full load of coal onboard, her freeboard was minimal. Thus the high seas could and did sweep aboard the ship.

WHEREAS A SHIP with twenty feet of freeboard will not take any water on deck in seas of fifteen feet, a ship with fifteen feet of freeboard will take every such sea aboard. This is not a problem as long as the deck itself is completely watertight. If none of the water that gets on deck finds its way below, the ship will remain afloat.

If there is an opening on deck that allows water to get below, every wave that comes aboard will spill down the opening. The ship will then sit a little lower in the sea (i.e., its freeboard will be reduced), and waves of slightly lower height can now come aboard. Because waves are passing by every few seconds, the amount of water that enters the ship increases every few seconds. With time, the ship will settle far enough that all the passing waves come aboard and spill into the ship's interior. Thus, the very act of sinking will increase the rate of sinking, and the process accelerates.

Hatches are openings in a deck. Until the advent of container ships, all ships carrying dry cargo had hatches so that most of the cargo could be stowed inside the ship. With small cargoes (such as boxes of tea in the nineteenth century), the hatches can be relatively small without impeding the efficiency of loading or unloading. Bigger cargoes require bigger hatches just to get the cargo inside. And to maximize the efficiency of loading of bulk cargoes (such as grain and ore), multiple large hatches, which may take up nearly all the deck space, are built into the decks of bulk carriers.

Hatches are, therefore, a potential source of leakage. Naval architects, shipbuilders, and especially sailors know it is imperative that hatches be constructed and maintained in a manner that keeps them watertight. Before the days of steel hatch covers, sailors made wooden hatch covers watertight by fitting layers of canvas and strips of wood (battens) over the hatch covers ("batten down the hatches"). With steel hatch covers, the canvas and wooden battens were replaced by waterproof gaskets set in the hatch covers, which would be bolted down tight. By the end of the twentieth century, the hatch covers on many bulk carriers had become immense and thus would be opened and closed hydraulically to ensure a watertight seal.

HELGA BOLTEN WAS a bulk ship; she was carrying coal. Thus, her hatches were large. If just one of those hatches were opened up, she could be subjected to serious flooding. Instead, two of her cargo hatches had been stove in by the seas, which allowed tons of water to rush below every time a sea came aboard the main deck. Furthermore, she had suffered some damage to her superstructure, so water was entering not just through the hatches but through the superstructure as well. Thus, she had three major sources of leakage.

There may have been still other sources for water entering the ship because *Helga Bolten*'s pumps were not able to control the flooding. She was slowly but inevitably sinking in the midst of a full gale on the North Atlantic Ocean in the impenetrable dark of an overcast night.

At 0207 on October 30, *Helga Bolten* sent an SOS on the international radio distress frequency. She was still under way and her pumps were still running, but her captain and crew believed they were losing the battle. *Helga Bolten* reported her position, the fact that the hatches were stove in, and pleaded that she needed "help as quick as possible."

One hundred thirty miles east of *Helga Bolten*, the *Casco*-class cutter *Chincoteague* was hove to right where she should have been: in position OS of Ocean Station Delta. *Chincoteague*'s crewmen were doing what they had done before on ocean stations and would do after on other ocean stations: making slow speed for hours steering west-northwest into gale force winds and thirty-foot seas.

Within five minutes of hearing *Helga Bolten*'s call for help, *Chincoteague* had increased speed and altered course to head for the stricken vessel. The good news was that *Chincoteague*'s best speed and *Helga Bolten*'s position meant the cutter could arrive on scene in less than nine hours. The bad news was that *Chincoteague* had to alter course fifteen degrees to head for the collier: The cutter would have to pound into the wind and seas the whole way.

Within two hours, *Chincoteague* was forced to slow down. Driving the cutter into the seas at just eleven knots was proving a challenge. She would crash into one wave after another, shuddering, shaking off the seas, and lurching ahead. The cutter and her crew had taken such abuse before and would take it again.

At some point, though, the captain has to make the decision to slow down to avoid damaging his own vessel. Stanchions smashed by seas or gear carried overboard by waves washing across the deck are the price that must sometimes be paid to save lives. But if *Chincoteague*, a sister ship to *McCulloch*, were to suffer deck or hull cracks while driving through these seas, she could become endangered herself. A broken rescue vessel is no rescue vessel at all.

Still, anyone on a rescue mission knows that minutes, even seconds can mean the difference between life and death. A decision to slow down is continuously questioned and reevaluated as the weather conditions are constantly reviewed. Any moderation in the weather, no matter how slight, will be seen as a sign to increase speed once more. If the cutter gives even the vaguest indication that she is riding more comfortably, it is an indication that it is time to push the speed up once more.

At 0405, *Chincoteague* once again increased her speed to 11.5 knots. Once again, the cutter was beat into submission, and thirteen minutes later, the speed was reduced to ten knots. *Chincoteague* plunged ahead through the predawn darkness, spray flying from the bow, spume washing over the superstructure, and seas inundating the foredeck.

Chincoteague was not the only ship to have heard *Helga Bolten*'s broadcast for help. The Cunard White Star liner *Mauretania* was about sixty miles northwest of the collier and immediately altered course for the German ship. *Mauretania* would be running down-sea and downwind toward *Helga Bolten* and was able to arrive on scene before 0600. A big, slab-sided liner, *Mauretania* was awkward to handle under adverse conditions and could not be maneuvered into position to complete a rescue, yet her crew stood by to recover *Helga Bolten*'s crew should the collier begin to take a plunge.

It was still quite dark. Sunrise would not occur until 0718, not that anyone could see it anyway; it was overcast and raining. *Chincoteague's* helmsman, steering by the intentionally dimmed light of the compass binnacle, struggled to see in the gloom, struggled to stay on his feet as the cutter pitched and rolled and shuddered ahead, and struggled with the wheel to keep on course as the ship corkscrewed through the seas.

A SHIP'S WHEEL, or helm, steers the ship via a series of connections (mechanical, electrical, hydraulic, or some combination of the three) that transforms the helmsman's input into action by the rudder itself. In the age of sail, the wheel was located aft, near the rudder, and was usually connected to the rudder either directly through gearing or indirectly by heavy line or wire that physically moved the rudder. As the size of ships grew, so did their rudders, and as ships became faster, the forces acting on those rudders increased tremendously. It is simply impossible, on a twentieth century or twenty-first century ship, for human beings to muster enough physical force to move or control a rudder. Instead, modern ships control the rudder through hydraulic rams, which not only apply more power but also can react more quickly thus giving the shiphandler better, quicker control over the vessel.

Oftentimes, the hydraulic rams are controlled by electric motors because it is more efficient to have an electrical signal sent from the helm (located forward in the ship) to the rudder space (located aft) than it is to have hydraulic piping running throughout the ship. Thus, modern ships have onboard a device called a steering motor, which reacts to the helmsman's inputs to move the rudder. The tops of the rudder posts, the rams, the steering motors, and all the electrical paraphernalia associated with the system are located in a space commonly referred to as the steering gear room or steering gear space. In fact, there is usually a duplicate steering wheel (often a reduced-size duplicate) in the steering gear room so that if some failure occurs in the system between the pilothouse and the steering gear room itself, the ship can still be steered.

At 0730 on October 30, *Chincoteague* had an electrical fire in her steering gear room. Had this fire not been brought under control at once, the result could have been catastrophic. An uncontrolled fire in the steering gear room could result in a complete loss of steering making it extremely difficult to control the cutter's direction. Without steering control, *Chincoteague* could soon find herself at the mercy of the wind and seas that had already damaged *Helga Bolten*. Even if *Chincoteague* could have maintained enough control by using the engines (i.e., steering by adjusting the speed of the port and starboard propellers) to proceed toward *Helga Bolten*, she would have been hard-pressed to help the stricken vessel once she arrived on scene.

Chincoteague's crew reacted promptly and properly, and the fire was extinguished. Unfortunately, the cutter could no longer be steered by the helm in the pilothouse. Fortunately, the helm in the steering gear room still worked. Now, *Chincoteague* would be steered not by someone peering out through the rain and spray into the unending seas but by someone buried in the bowels of the cutter who could see nothing but a gyrocompass repeater and a maze of pipes, wires, machinery, and exposed ship structure.

The watch officer had to remain on the bridge to keep a proper lookout and conduct his navigation. Therefore the helmsman, now located in the steering gear room a couple of hundred feet away, could not get his orders directly from the watch officer but had to have them relayed to him by a phone. It was like flying on instruments and taking directions from an air-traffic controller over the radio for an aircraft pilot. It was like driving with the windshield painted black and instructions coming in over a cell phone to the driver in a car. *Chincoteague* pressed on.

An hour later, *Chincoteague* increased speed again to eleven knots. Another hour later and she was up to 12.5 knots. *Mauretania* gave *Chincoteague* updated positions on the *Helga Bolten*, and these showed that the cutter should change course more to the west and then west-southwest, which meant she was pounding less (but rolling more) allowing further increases in speed. By 1030, *Chincoteague* was up to fifteen knots.

Thirty minutes later, *Helga Bolten*'s captain notified *Mauretania* that it was impossible for the ship's crew to make repairs, that he doubted the ship's ability to make port, and that he was intending to abandon ship. The United States Coast Guard was getting close, however, so *Helga Bolten*'s crew stayed aboard their ship.

SHORTLY AFTER 1300, *Chincoteague* exited the western edge of Delta, still steering west-southwest. Within an hour she had *Helga Bolten* and the *Mauretania* in sight. By 1500, *Chincoteague* was on scene and, with the concurrence of *Helga Bolten*'s captain, *Mauretania* was released to resume her voyage. *Helga Bolten* sent a radio message to *Chincoteague* stating that the German ship would not stay afloat until the next morning and that her boats could not be launched because of the list and sea conditions and finally asking that the crew be taken off as soon as possible. All *Chincoteague* had to do was remove thirty-three men from a listing, sinking ship in rain, 35-knots of wind, and 30-foot seas.

Chincoteague's captain, Cdr. Raymond G. Miller, decided that using the cutter's small boats to pick up the survivors was too risky: The high seas, strong winds, and inevitable battering the boats would take while alongside both vessels seemed a recipe for disaster. Instead, *Chincoteague* would pass a line to the collier and use that line to run rafts between the two vessels.

Coast Guard cutters and many other ships, then and now, carry guns that are designed for and equipped to shoot a line from one ship to another. These are not big, complex, deck-mounted weapons but rather simple, smooth-bore, shoulder-braced firearms. Instead of bullets, these guns fire a brass (or other) projectile that is designed to have a very light piece of line or cord attached to it. That piece of cord is attached to a somewhat heavier line, which is subsequently attached to an even heavier line, until eventually a line sufficiently strong to pull a raft through the water is run between the two ships.

There are two challenges to passing the first, lightweight line: The ships have to be close together and the projectile has to pass over the receiving ship so that the line can be retrieved. This is not especially

Two rafts holding *Helga Bolten's* crew are pulled to *Chincoteague.*

difficult in sheltered water or even in open water when the wind and seas are moderate. *Chincoteague*, however, was in the middle of the North Atlantic in a full gale. Furthermore, daylight was running out. Sunset was at 1750, and the overcast guaranteed it would be dark shortly thereafter.

Captain Miller nosed the cutter up close to *Helga Bolten*, the gunner's mate fired his weapon, and the projectile hurtled toward the ship. *Chincoteague* repeated this evolution five times and failed five times because the wind would carry the projectile away from the ship, or the projectile would fall short, or the lightweight line would take up too much strain as the vessels rolled and pitched together and apart, or the seas would simply snap the line.

The gunner's mate had only one projectile left, one final chance to do what had to be done. It was 1640 and darkness was closing. *Chincoteague* approached, the gunner's mate fired, the line held, the heavier line was passed and it held, and then the heaviest line was passed. Two fifteen-man life rafts were attached to the line, and *Helga Bolten*'s crew hauled them over. They clambered into the rafts, and *Chincoteague*'s crew hauled them back and got them alongside the cutter by 1752. *Chincoteague* stopped, and the cutter drifted beam to the wind and seas with the rafts resting along the leeward side. There they were safe from wind-induced capsize and felt minimal motion in comparison with that of the cutter.

BEING IN A SMALL CRAFT, either a boat or a raft, alongside a ship is like being next to a steel wall. The ship's side is absolutely vertical and perfectly flush with no ladders, ramps, handrails, or other way to get from the water's edge up to the main deck. It is impossible to get out of a small craft and onto a ship without some means of vertical ascent.

Therefore, Coast Guard cutters carried (and still carry) devices officially known as embarkation nets but commonly referred to as scramble nets, which can be quickly lowered over the side. These are the same type of nets one sees in vintage war films when assault troops go from their ship to their landing craft. They are indeed nets: pieces of line tied together in a pattern of squares, and when the net hangs down the side of the ship, it provides minimal toe and foot holds so that one can climb up the side of the ship to safety.

Or at least one can climb up the side to safety if one is young, in good physical condition, and well rested. Unlike *Chincoteague*'s crew, most of *Helga Bolten*'s merchant crew was not uniformly young and limber. Furthermore, they had just spent most of the day on a ship sinking in a gale, and they had just had a harrowing ride in rafts through thirty-foot seas. Some were tired, some were out of shape, and many of them simply could not climb the nets.

Four of *Chincoteague*'s crew, Ens. Brinton Shannon, Seamen Harmon Schreiner and Robert Hayman, and Fireman Gilbert Wagner, volunteered to climb down the nets and help *Helga Bolten*'s crew. It was dark, and though the wind had eased to twenty knots, the seas were still running at about twenty feet, so *Chincoteague* was rolling deeply. And these four young men offered to step off their own main deck and struggle down the awkward net to help thirty-three strangers to safety.

Helga Bolten listing but still afloat.

They took lines with them to fasten around *Helga Bolten*'s crewmen so that the rest of *Chincoteague*'s crew could pull the beleaguered survivors up on deck. Pushing, pulling, tying lines, giving directions, slipping, falling, and finally struggling up the nets themselves, these four men got all the survivors aboard.

By 1815 they were finished. In less than three hours after she first arrived on scene, *Chincoteague* had shot the lines, passed the rafts, got them alongside, and wrestled thirty-three weary and wet survivors aboard. No one was even seriously injured. It was a "textbook" rescue.

A FULL GALE, a sinking ship, the entire crew saved without so much as a serious injury: It sounded like mission accomplished and case closed.

The sinking ship, however, stopped sinking. As the weather continued to moderate and as waves no longer swept *Helga Bolten*'s decks, she stopped taking on water. *Chincoteague* had remained close aboard the powerless and therefore unlit vessel on October 31 to warn away other ships.

On the morning of November 1, *Helga Bolten*'s captain, Albert Ewart, looked across the ocean at his ship. With winds reduced to ten knots, the seas down to less than five feet, and the light of day rather than the gloom of night facing them, he and his third engineer, Martin Prigg, thought they could return to their ship, evaluate her condition, perhaps even make some repairs, pump out the water, and possibly get under way again.

Chincoteague's Captain Miller understood the devotion to ship and duty that Captain Ewart felt. He agreed to transport the two men back to *Helga Bolten* using one of *Chincoteague's* motorboats. Though this was not an act of bravery of the sort *Chincoteague's* crew had displayed in getting *Helga Bolten's* crew off, it was an act of generosity and selflessness. Captain Miller had risked his crew and his cutter to bring these men to safety, and he now agreed to undo what he had done just thirty-nine hours earlier.

The motorboat carried the two officers back to their ship at 0900. They found eleven feet of water in the engine room but no appreciable deterioration in the ship's condition. She had a twenty-degree list to port, and Captain Ewart estimated that, if nothing further was done to the ship, she would stay afloat for two more days.

Meanwhile, the German ship's owners had arranged for a salvage tug to proceed to *Helga Bolten*. Captain Ewart and Mr. Prigg decided to remain aboard their ship to help with the salvage effort once the tug arrived on scene.

In the early afternoon of November 2, the tug *Ebro* arrived hoping to dewater the *Helga Bolten* and take her under tow. *Chincoteague's* small boat shuttled *Helga Bolten* crewmembers from their own ship to the stricken vessel and then back again and also transferred *Ebro* crewmembers and a gasoline-powered pump from the tug to the collier. The weather remained fair with light winds and low seas.

By the evening of November 3, *Helga Bolten's* engine room was nearly pumped out, the forward hatch was almost watertight, the list was decreasing, and the ship's condition seemed hopeful for the first time in four days. The weather was still benign.

Throughout all this, the German ship and the U.S. Coast Guard cutter had drifted back onto and around the confines of Ocean Station Delta. *Chincoteague* was not only executing the search-and-rescue case but also keeping station as well.

On November 5, a different salvage tug, the *Ocean*, arrived to relieve *Ebro*. Once again, *Chincoteague's* small boat began shuttling personnel. Captain Ewart and Mr. Prigg were joined by five *Helga*

Bolten crewmembers who returned to the ship to secure the second hatch cover and make the necessary preparations to take *Helga Bolten* under tow. *Chincoteague* delivered them to their ship along with a fifteen-man raft.

Then the inevitable happened. It was November in the North Atlantic, and the wind began to blow and the seas began to build. And the waves began to wash over *Helga Bolten*'s decks. Her crew could not close up the second hatch. She started taking on water yet again. *Helga Bolten*'s captain, her third engineer, and five crewmembers were finished; they climbed into the raft and paddled away from their stricken ship, though four members of *Ebro*'s salvage crew remained aboard. *Helga Bolten*'s crew had wisely left a line attached between the raft and the collier so that the *Ebro* crewmembers could pull it back to the collier if they needed it. *Chincoteague* managed, once again, to get all seven out of the raft and onto the cutter's deck.

The pump that had been keeping *Helga Bolten* afloat ran out of gasoline and stopped. The weather made it impossible for *Chincoteague* to pass more fuel over. The water rose inside *Helga Bolten*. Now *Ebro*'s crewmembers were finished. They clambered into the raft and were picked up by *Ocean*.

On the morning of November 7, the *Casco*-class cutter *Barataria* arrived at position OS on Delta and relieved *Chincoteague* of her ocean station duties. *Barataria* did not relieve *Chincoteague* of *Helga Bolten*'s crew nor of her obligations to assist with the collier's salvage. By that afternoon, the weather had moderated once again, and *Ocean* put a salvage crew back on *Helga Bolten*, *Chincoteague* transferred cans of gasoline to the collier, and *Ocean* told the cutter that no further assistance was needed as the tug prepared to tow the collier toward the Azores.

Chincoteague headed for home—Norfolk, Virginia—last seeing *Helga Bolten* down by the head and with a fifteen-degree list. This time it *was* case closed with all survivors rescued successfully, some of them twice.

RESCUE CASES WERE and are part of the Coast Guard's regular duties. Rescue cases usually occur in nasty weather because it is the weather that creates the problem causing the need for rescue. But *Helga Bolten* was an exceptional case and therefore impressed the service's senior officers. The commandant of the Coast Guard, Vice Adm. Alfred C. Richmond, told Captain Miller and *Chincoteague*'s crew that their performance was "in the highest tradition of the Coast Guard" particularly as they had completed the rescue "under most hazardous weather and sea conditions."

The government of then–West Germany also was impressed, not just with the initial successful rescue of *Helga Bolten*'s crew by *Chincoteague*'s crew but also with their tenacity in standing by the stricken vessel, moving people, pumps, and fuel around, and picking up survivors for a second time. To show the government's appreciation, in August 1957 the West German charge d'affaires presented a bronze plaque to the cutter's officers and crew at a ceremony in Washington, D.C., thanking them for their assistance and praising their seamanship.

Thereafter, *Chincoteague* went back on ocean station.

A SHIP CAN BE DAMAGED by a single strike: an iceberg, a huge wave. Or it can be damaged by a series of smaller blows each of which individually would not be a problem but that collectively creates catastrophe. *McCulloch* on Bravo in 1959, for instance, was pounded day after day by high winds and heavy seas from a single, unmoving storm. Somewhat differently, *Owasco* on Bravo in 1962 struggled through a series of brief, intense storms that, over the course of several days, forced her to abandon station.

Whereas the cutters on ocean stations expected to suffer through adverse weather, they did not expect to be surprised by the onset of storms whether the storms were weak or strong, short or long. Contrariwise, they expected to be the beneficiaries of weather forecasts because they were providing the real-time weather information that those ashore used to produce accurate forecasts.

The *Casco*-class cutter *Half Moon* was on Ocean Station Dog (later Delta) on October 3, 1949. The weather was so favorable (ten to twelve knots of southwest wind, slight seas, and eight miles of visibility) that the cutter completed a scheduled gunnery exercise in the early afternoon. There was nothing in the forecast to indicate any change in conditions.

However, within thirty minutes after the gunnery exercise, the wind was blowing from the north at more than forty knots, and the seas started to build. An hour later, the wind had increased to at least sixty knots, and the seas were quickly building to more than thirty-five feet.

Half Moon's captain, Cdr. Gilbert I. Lynch, decided to head into the seas. His decision was based in part on his desire to keep his cutter on station. He knew that if he ran with the storm instead, the cutter's speed would combine with the wind and waves to soon push *Half Moon* off station to the south. By 1500, *Half Moon* was using both propellers as well as her rudder to keep herself heading into the seas.

A SHIP'S DIRECTION, or heading, is mostly controlled by its rudder. More precisely, its direction is controlled by water moving past the rudder. When a ship is stopped and there is no current, the rudder will have absolutely no effect on which way the ship faces. It is analogous to turning the front wheels on a parked car.

Once water starts to move by the rudder, however, the rudder does have an effect on the ship's heading, even to a limited extent for a ship at anchor. If water moves by a rudder more quickly, the effect becomes greater. Thus, by increasing speed, *Half Moon*'s crew expected to increase the rudder's effectiveness and increase the cutter's ability to hold her course into the wind and seas.

Ships with multiple propellers have those propellers offset from the centerline. Therefore, as one propeller turns, it not only pushes the ship ahead but also has a tendency to push the ship toward the side opposite the propeller: Going ahead on the starboard shaft pushes the ship's bow toward port, and going ahead on the port shaft pushes the ship's bow toward starboard. By increasing (or decreasing) speed on one propeller, the shiphandler can augment the rudder in changing the ship's direction when rudder action alone is insufficient to change course. By using both propellers at the same speed, *Half Moon*'s crew expected to acquire a balanced thrust.

Despite the fact that she was heading into the seas, which should have given maximum rudder effectiveness, *Half Moon*'s rudder became almost useless. In addition, slow speed for both propellers was insufficient to hold course: Even with both shafts turning for two-thirds ahead, which should have been about ten knots of speed, the crew often found it necessary to increase one shaft to standard speed ahead to hold course.

This was happening to *Half Moon* for two reasons. First, every slight variation in the wind's direction (and the wind does in fact constantly shift direction ever so slightly) would force the bow off course. Second, as the wind continued to blow, it built up the seas, and each wave also pushed the bow around. Consequently by late afternoon, *Half Moon* was struggling to hold her course in the ever-building wind and seas. Worse, there was no indication that the weather was abating. Without an accurate forecast, *Half Moon*'s captain chose to continue heading into the seas.

A SHIP FACED WITH truly adverse weather has two options: run with the storm or heave to (i.e., head into the wind and waves). The best course of action is driven by a number of factors: (1) What is the ship's mission? (2) How adverse is the weather? (3) How adverse is the forecast? (4) How near is danger (i.e., a rocky coastline; shallows)? (5) How is the ship loaded: filled with cargo or empty? (6) How has the ship handled such weather in the past?

There are advantages and disadvantages to each option. The ship will ride more comfortably with less stress on both ship and crew when heading down-sea and downwind (i.e., running with the storm). However, the ship will be under more control ("more" being a relative term) heading into the wind and seas (i.e., hove to).

Running with the storm carries the danger of having the following seas overrun the ship with the result that huge masses of water may bury the stern, potentially sinking the ship stern-first. This phenomenon is known as pooping, from the name for the aftermost, or poop, deck. At the least, a following sea will push the stern around, minimizing the rudder's effectiveness and causing decreased control.

Heaving to, on the other hand, means the ship will pound, even pound severely, with the threat of huge waves burying the bow as happened to *McCulloch* in 1959.

In modern ships, heaving to is often the preferred technique in absolute survival conditions. The reason why ships head into the seas as a means of survival is related to ship design and ship movement.

A ship design is said to be a series of compromises. A ship must be large enough to carry its cargo, but if quantity of cargo were the only concern, then the ship would be huge.

The speed with which that cargo must be delivered will also play a part in the ship's design. Bulk and speed are generally incompatible. For instance, the great clipper ships of the mid-nineteenth century carried comparatively little cargo (people, tea), but it was cargo that paid a premium for a fast passage. The result was that the hulls of those ships were narrow, fine-ended, and shallow when compared with the hulls of their contemporaries that carried wool or wood.

Ships move through two fluids (air and water) and encounter both air resistance and water resistance. Because ships constantly push against water while under way, they typically have narrow bows to minimize the water's resistance. But something that is narrow is more susceptible to deformation than is something that is flat (think of pushing on the end of a piece of cardboard versus pushing on the end of a cardboard box). Consequently, ships often have more internal structural members designed in at the forward end than elsewhere in the hull in part to help minimize this deformation.

Furthermore, when ships strike objects (piers, bridges, other ships), they usually do so with their bows. This is the inevitable consequence of the fact that ships are, most of the time, moving forward rather than backward. Thus, the extra strength built into a ship's bow provides protection against two problems: (1) the routine daily pressure of the water on the bow and (2) the far less routine striking of the bow against solid things.

Heavy seas are a cross between routine water pressure and striking something solid. Certainly a big wave is water. But water is nearly incompressible, and so a big sea has some of the properties of a solid object. It only makes sense that if a ship is going to punch its way through heavy seas, then it will have a strengthened bow to do so.

Strength is only part of the story; control of the ship is the rest. If the ship is moving with the waves at exactly the same speed as the waves, the rudder will have no effect. It is rare that a ship will sail down-sea at exactly the same speed as the waves, nevertheless the rudder's effect is reduced when the ship is sailing with, rather than into, the waves.

Thus a ship that is heaving to is going to do so by heading into the waves to maximize control and put the strongest part of the ship into the seas. If the ship heaves to at minimal speed, there is minimal pounding and a minimum number of heavy seas breaking aboard. At the same time, the minimal ship speed is augmented by the speed of the waves (which are moving in the opposite direction) to maximize rudder control. Ships have done this for centuries because it works.

THE CREW OF *Half Moon* expected this to work, but the late-afternoon hours were catastrophic. The wind quickly increased until the cutter's rigging began to howl, and the seas built to tremendous heights. Captain Lynch reported wave heights of sixty to seventy-five feet. Seas of this magnitude are often described as mountainous, but in this case that term was improper. The distance between these seas was too short for them be seen as rolling mountains. They were great, gray walls of water so close together that *Half Moon* could not slip up the face and slide down the back of the waves. Instead, as she struggled up and over the breaking crest of each wave, the cutter's bow would emerge into open air where it would hang for a few sickening moments before gravity slammed it down into the ocean only to be crushed by the next wave.

Half Moon pounded into these brutally steep seas over and over. This was not just the occasional teeth-jarring crash that is part of life at sea. These were repeated, crushing blows to a cutter struggling to stay afloat. The violence of the motion began to tear things loose that were considered immovable.

It was and still is a common practice aboard ships to keep standard-size oxygen and acetylene tanks (necessary for performing maintenance and repairs) stored upright somewhere on deck. These tanks are steel cylinders, about a foot in diameter, and about five feet tall. To keep these tanks where they belong, they are routinely held in place by solid steel bars or shaped steel straps, which are themselves held in place by heavy bolts.

It was and still is a common practice aboard ships to store spare parts on shelves. Because all ships are subjected to constant motion, nothing would stay on the shelves if they were completely open.

Because a ship at sea has to act as its own repair shop, many of these parts are heavy engineering pieces such as pumps and pistons and big electrical motors. Consequently, these shelves not only have lips on their edges but also have multiple horizontal steel bars, known as battens, which run across the openings of the shelves several inches above the bottoms of the shelves to hold the heavy parts in place.

Half Moon pounded so hard and so long that all the oxygen and acetylene tanks came adrift, working themselves free from the heavy, bolted brackets. These tanks then rolled and crashed around the cutter, damaging bulkheads, rails, and anything in their path. *Half Moon* pounded so hard and so long that heavy boxes of spare parts jumped over the tops of the battens and crashed onto the deck.

Half Moon pounded so hard that some of her internal structural hull components, big pieces of steel designed to give the cutter her strength, twisted and broke. She pounded so hard that one of her internal bulkheads permanently warped to the point that the watertight door in it could not be closed. She pounded so hard that much of the steel-plating on the hull bottom was dished in. This is nearly unbelievable: Damage of this severity usually occurs only when a ship has struck something solid such as a rock, a pier, or another ship.

Half Moon pounded so hard that some of the steel plating of the hull bottom fractured. This broken plating let water inside. In other words, *Half Moon* started to sink.

Fortunately, *Half Moon* kept electrical power, which kept the pumps running, thus the flooding could be kept under control. Equally fortunate, as the storm continued to blow, the distance between the seas began to increase even though their size did not decrease, and *Half Moon* started to ride more easily so that the pounding diminished. It also took less power to keep the cutter on course, so she was slowed a little, and this further eased the pounding.

By the next morning, the wind had "diminished" to forty-five knots, and the seas also began to decrease. The morning after that, forty hours after the onset of the storm, the weather had settled sufficiently that *Half Moon* could struggle back to the center of station. She was relieved four days later and headed to New York where, fifteen days after the storm ended, she was in dry dock having all the damage repaired.

*   *   *

HALF MOON'S DAMAGE was reported up the Coast Guard's chain of command. Ten years later, *McCulloch,* also a *Casco*-class cutter, would have her deck start to crack on Ocean Station Bravo. That damage, too, was reported up the chain of command. The crews of other cutters of the class had also reported cracks whose only apparent source was nasty weather. Those who sailed this class of cutter were, justifiably, concerned that their cutters might literally break up at sea solely because of weather damage.

As *McCulloch*'s report worked its way up the bureaucracy in 1959, it became obvious that many of the cutters of this class were suffering damage beyond "normal." Any seaman could understand waves that came aboard and carried away boats. Anyone who had sailed in cold weather could understand icing severe enough to break stanchions or shatter antennae. But these cutters were having hull failures—structural damage that should never have been caused simply because the ship was "working" too hard in adverse weather.

Why, then, did this cracking happen to the *Casco*-class cutters and not to those of the *Secretary* class or *Owasco* class, which were their contemporaries? There are two possible answers, one related to design and one related to the number of cutters.

The *Casco*-class cutters were built for the Navy for use as seaplane tenders during World War II. They were expected to spend most of their time tending seaplanes, which is to say they would be anchored and operating in comparatively sheltered waters where seaplanes could operate with relative safety. Their design included a wide, open fantail where seaplanes or heavy seaplane parts could be hoisted aboard for repair or maintenance. They had huge hatches forward on the 01-deck, the main deck, the first deck, and so forth right on down to the magazines (ammunition storage spaces) so that bombs for the seaplanes could be easily stowed and loaded.

Furthermore, these vessels carried an incredible amount of fuel for their size. The tanks that had been designed and built to carry aviation fuel were simply converted for use with diesel fuel when the ships were transferred to the Coast Guard. This gave the *Casco*-class cutters "long legs" and high endurance, necessary attributes for patrols on the ocean stations.

Casco class: The cutter *Humboldt*.

Tank conversion was not the only change made to these vessels. When built, they had very large cranes located aft, again necessary to work on seaplanes. Once the Coast Guard took possession of the ships, the cranes were removed because they served no real purpose on ocean station. Simultaneously, the Coast Guard added balloon shelters—large enclosed spaces in which weather balloons could be inflated prior to release—at the aft end of the superstructure.

Other things required no change. The ships of the *Casco* class had been designed and built during the war, and thus they had great redundancy. There were two completely separate engine rooms, each one of which had two main engines to drive one of the propellers as well as generators for ship electrical power. Thus one engine room could be completely destroyed and the ship could still remain operational.

This was all part of the design compromise: These ships had to have open working spaces and big cranes. They had to have big openings in the deck for bomb handling. They had to have redundancy in case of combat damage. They had to have huge fuel capacity to refuel seaplanes.

The foredecks were flush, that is to say there were no bulwarks (low bulkheads) to help deflect spray and seas as these struck the bow because it was not expected that the vessels would spend days or weeks pounding through the worst of the winter North Atlantic. As a result, the designers of the *Casco* class, though not ignoring the very real possibility that the ships could encounter heavy seas, would view that possibility as remote rather than routine. It would be worked into the design as a worst-case circumstance, not as an expected operational standard. Thus, the ships were designed and built with comparatively thin steel and small internal structures. This is known as light scantlings.

The *Secretary*-class cutters, on the other hand, were designed with the intent that they go far offshore in all weather and stay there as a matter of course. They had bulwarks forward, and they were narrow at both bow and stern. And, though the *Owasco*-class cutters lacked the raised forecastle of the *Secretary* class, they, too, had been designed and heavily built to work as offshore escorts. Both of those classes (*Secre-*

tary and *Owasco*) *were* expected to face the worst weather and face it consistently and repeatedly.

Thus design itself can explain some of the damage to the *Casco*-class cutters.

A further explanation comes from the fact that the *Casco*-class cutters outnumbered the other ocean station cutters. There were eighteen *Casco*-class cutters in the Coast Guard's fleet. There were thirteen *Owasco*-class cutters, and one of those, *Iroquois*, was taken out of active service in mid-1954 leaving only twelve thereafter. There were six *Secretary*-class cutters. In other words, from 1954 until the late 1960s when the *Hamilton*-class cutters started to be commissioned, the *Casco* class alone made up half of the ocean station fleet.

Inevitably, the greatest number of cutters would accumulate the greatest number of days at sea. The more time a ship or group of ships spends at sea, the more time it will be exposed to adverse conditions. The numbers nearly guaranteed that the *Casco*-class cutters would collectively spend more time exposed to adverse weather than would the other classes.

This also explains why a disproportionately high number of the *Casco*-class cutters became involved in spectacular search-and-rescue cases. They were simply in the right place at the right time because the numbers made it nearly inevitable.

* * *

McCulloch's 1959 REPORT OF DAMAGE was answered in an official letter written by the commandant of the Coast Guard, Vice Adm. Alfred C. Richmond.[1] The commandant pointed out the severity of the storm, the short length and great height of the seas, and concluded that *McCulloch*'s topside damage was "not considered undue."

He also noted that some of *McCulloch*'s deck fractures had occurred at the site of welds or other "structural discontinuities," that none was more than a few inches long, and that some of them were "of former origin." He did not describe what that "former origin" might have been, thus leaving open a question if those cracks, too, had occurred because of weather.

The commandant proceeded to address the fact that *McCulloch* and *Half Moon* were not alone among the *Casco*-class cutters in suffering internal (structural) damage. He pointed out that these cutters were built of light scantlings and that therefore only a complete redesign and rebuild could eliminate all these cracking problems. He observed that these cutters had "a long record of tough duty" and that none had suffered "major" fractures.

In the end, the commandant's analysis was correct: After her brutal patrol on Bravo in 1959, *McCulloch* would serve for another thirteen years (*Half Moon* served for another ten years after her brutal 1949 patrol). None of the *Casco*-class cutters ever broke up catastrophically despite three decades of storms and abusive weather.

Still, the commandant's letter contained this phrase: ". . . although there is a possibility of a major failure just as there is always a possibility of failure in any light scantling highly stressed structure, the probability of a catastrophic failure is not considered of consequence . . . ." That left the *Casco*-class cutterman to ask himself, "Not of consequence to whom?"

Afterward, that same cutterman on ocean station would be constantly recalculating the "probability of a catastrophic failure" as he looked up and up and up, thirty, forty, or fifty feet up, to the top of the next wave as it readied itself to crash down on his *Casco*-class cutter.

ENDNOTE

1. Richmond (1959).

6

ECHO

35°00′ N, 48°00′ W

OCEAN STATION ECHO had a relatively benign reputation when compared with those of Bravo, Charlie, and even Delta. Air temperatures never got as cold as they did on the more northerly stations, and thus there was very little threat of icing up. Water temperatures, too, stayed well above freezing, which not only helped eliminate the icing problem but also gave individuals in the water a better chance of survival.

Ocean Station Echo was not without threats, however. During hurricane season in the North Atlantic, the station was susceptible to nature's most violent storms as witnessed by *Absecon* and *Pamir* in 1957. Nor were midwinter storms absent. Wintertime North Atlantic low-pressure systems may not have been as deep (low barometric pressure) or as strong (wind force and consequent seas) on Echo as they were farther north, but they could still wreak havoc.

On January 10, 1966, the grain-carrying ship *Monte Palomares* (8,970 tons, 475 feet) was hove to in more than fifty knots of wind and thirty-foot seas just north of the limits of Ocean Station Echo. *Monte Palomares'* problems began at 0600 when the cargo of bulk grain shifted in the heavy seas giving the ship a fifteen-degree list. Four hours later, the main propulsion engine broke down, and the crewmen of *Monte Palomares* were left with no option except the one forced on them and their ship: lay beam to the wind and seas. This apparently caused further shifting of the cargo and aggravated the list.

During the afternoon as the ship rolled deeper to starboard, the vents for the ballast tanks began to submerge in the waves, and water ran down the vents and began to fill the tanks. This water made the list even worse. Without power, with a shifted cargo, and with the ship taking on water through the ballast tank vents, it seemed the end was nigh for *Monte Palomares*. *Monte Palomares'* radioman broadcast a distress message reporting a thirty-degree list and that the crew expected to abandon ship soon.

Meanwhile, the *Owasco*-class cutter *Escanaba* had been hove to on Echo. After her captain, Cdr. Robert E. Foley, heard *Monte Palomares'* call for help in the late afternoon, he altered course and increased speed bound for the stricken freighter and expected to arrive by 1000 the next morning.

Escanaba was not the only ship to receive *Monte Palomares'* distress message nor was she the only one to respond. *Steel Maker*, a 7,700-ton cargo ship, headed for the bulk carrier as did the *Thuro Maersk*, a 5,900-ton Danish ship.

No one would get there on time. *Monte Palomares'* captain ordered the crew to abandon their ship at 1730, an hour before sunset. The crew had obviously and prudently stayed with their ship as long as possible—*Monte Palomares* sank shortly after everyone abandoned.

The ship's list meant the port lifeboat could not be lowered. Consequently, the crew lowered the starboard boat and launched two rafts. They were now at the mercy of the sea in the dark North Atlantic night.

At 2057, *Thuro Maersk* arrived at the sunken ship's last known position, and her crew started searching. The Coast Guard had sent a C-130 airplane out of Argentia, Newfoundland, to search as well. The airplane arrived an hour after *Thuro Maersk*.

Nighttime searches are extremely difficult. There is very little chance that the crew of an airplane or a ship will see a boat or a raft, let alone an individual person in the water. The hope is that survivors will shoot flares or have some sort of light with them so that they can be seen. The crews of *Thuro Maersk* and the C-130, however, saw nothing. More ships would join the search, and the Coast Guard C-130 would be relieved by a U.S. Air Force plane.

The crew of *Steel Maker* expected to arrive on scene at 0630 on January 11. But at 0230, while still sixty miles east of the last known position of the grain ship, *Steel Maker* came across a raft with four of *Monte Palomares*' crew in it. Somebody's navigation was wrong.

THE CREW ABOARD *Steel Maker* asked the men they had rescued what the four knew about the rest of *Monte Palomares*' crew. These survivors reported that eighteen men were in a lifeboat and five were in another raft. They also said they had watched the captain and "many" other crewmembers drown as they all abandoned ship.

This type of information can be useful, even necessary for those conducting a search because they have to know what, exactly, it is they are searching for. A lifeboat is usually larger than a life raft and is thus easier to see. That means searching units can be spread a little farther apart when looking for a boat but should be drawn more closely together when looking for a raft. A raft will usually drift faster than a boat, and so a search area will have to be expanded to find both a boat and a raft. If the searchers know exactly how many people are missing, they will be able to call off the search once all of them are accounted for.

However, it is nearly inevitable that information obtained from survivors will be incomplete at best and inaccurate at worst. *Monte Palomares*' four survivors said they saw "many" of the crew drown. How many is "many?" Without knowing that, the searchers did not know how many they were searching for. *Monte Palomares*' survivors reported eighteen men in the lifeboat. But how accurate could their counting be when everyone was struggling to leave a listing, sinking ship during a howling gale near nightfall? In fact, the number of men reported to be in the lifeboat was later changed to "fifteen to seventeen" and later still to "sixteen."

Compounding these uncertainties were the inevitable discrepancies arising from translation (*Monte Palomares*' crew was Spanish; *Steel Maker*'s was American), radio transmission errors or misunderstanding, and use of apparently simple but contradictory terms. For instance, the crew of *Steel Maker* first reported that they had picked the four survivors up from a "small boat." In fact, the four survivors were in a raft.

None of that stops a search effort. Searchers keep searching even as they try to resolve the discrepancies in reports and thereby figure out exactly what and who they are still searching for. So the Coast Guard C-130 kept flying and looking while the crews of *Escanaba, Steel Maker, Thuro Maersk*, and the other ships kept sailing and looking.

Steel Maker's recovery of survivors provided one other piece of information that is critical to search planning: It gave a *datum*. For search planners, datum is the point of origin for starting a search. It may be and often is the location of the sinking. But if that position is uncertain as it was in the case of *Monte Palomares*, then locating a survivor will provide a datum even though it is not a perfect one. It will be rare that the survivors are still at the position where the vessel sank because wind and currents will have pushed them away. But the wind and currents will have similar effects on all other survivors, and thus finding one survivor gives a very good approximation of where the rest may be found.

THUS THE COAST GUARD C-130 accurately established *Steel Maker*'s position at the time her crew located the raft thereby giving a new datum and enabling better computations for the estimated drift of the raft. The result shifted the search area 120 miles east-southeast of the original datum. *Escanaba* headed for the new search area.

At 1400 on January 11, *Escanaba* started a coordinated search with a U.S. Air Force plane using the new datum and search areas. Fifteen minutes later, the Air Force aircrew spotted what it thought was a raft with three people in it. *Escanaba* headed for the raft at full speed, and within an hour her crew had the raft in sight. Thirty minutes after the aircrew sighted the raft, *Escanaba* had it alongside.

That was the easy part; the hard part would be getting those men aboard the cutter while the seas continued to run at twenty-five feet. *Escanaba*'s crew lowered the scramble net, the swimmers harnessed up, and an extra line was rigged out from a boat davit. The swimmers

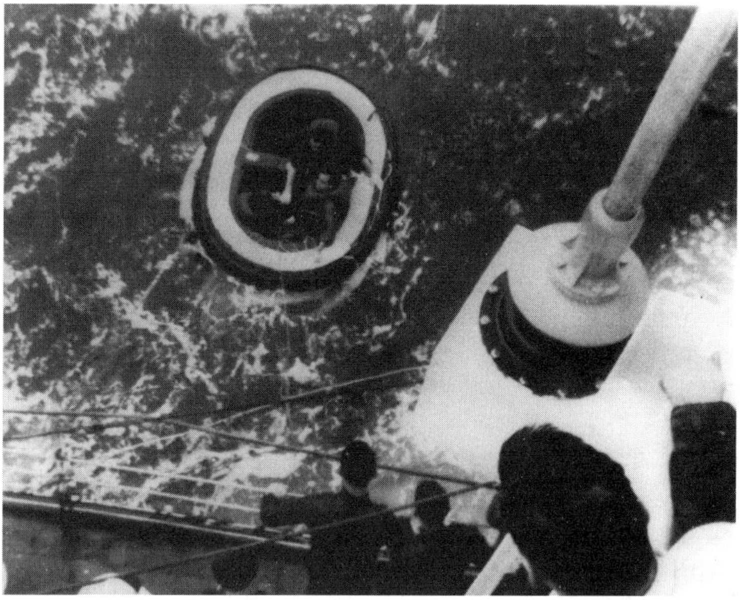

Two Coast Guard volunteer swimmers stand by on the deck of *Escanaba* to recover two survivors and one body from the *Monte Palomares* raft.

descended the net, hooked the survivors to the line, and helped hoist them on deck. Although it did not take long, Captain Foley would describe the evolution as "hair-raising but successful."

That is to say it was successful in that all three got aboard. One of the men, however, was dead even as *Escanaba* came alongside the raft. The other two reported that their day in the life raft had been a rough one: The heavy breaking seas had capsized the raft ten different times before the men were rescued by the crew of *Escanaba*.

Seven of *Monte Palomares*' crew of thirty-eight were now accounted for. There were still thirty-one missing. The search would go on.

The following day, January 12, the crew of *Escanaba* would find a few empty life jackets and some steel drums, which the survivors said were part of *Monte Palomares*' deck cargo. The crew also found a pair of lifeboat oars and lifeboat provision boxes. In light of those discoveries and the number of times the raft of the two survivors had capsized, it seemed likely that the lifeboat from *Monte Palomares* had either capsized or sunk. People in the water in January in the North Atlantic have a life expectancy measured in minutes. The crew of *Monte Palomares* had abandoned ship two days earlier. There were still thirty-one missing. The search would go on.

The *Casco*-class cutter *Barataria* had been dispatched from the Bermuda SAR Standby to aid in the search-and-rescue mission. On the morning of January 13, *Barataria* arrived and relieved *Escanaba* as the on-scene commander. *Escanaba* headed directly for Argentia to off-load the survivors and the deceased. The merchant ships were released to continue their voyages. *Barataria* would continue to sweep the area until the morning of January 14 when the Coast Guard's Commander Eastern Area would suspend the search and send *Barataria* back to the Bermuda SAR Standby.

There are still thirty-one missing.

THE SAME STORM that sent *Monte Palomares* to the bottom began breaking up the steamship *Lampsis* on January 10, 1966, some 240 miles southwest of the site of *Monte Palomares'* sinking. Less than an hour before *Monte Palomares* was abandoned, *Lampsis'* captain, Stefopulous Dimitrious, ordered his radioman to send out an SOS. At 1658, the *Owasco*-class cutter *Androscoggin* heard the call for help. *Androscoggin* had originally been on her way to relieve *Escanaba* on Ocean Station Echo, but both cutters had been diverted to the *Monte Palomares*. Because *Androscoggin* was closer to *Lampsis* than was *Escanaba*, the Coast Guard's Commander Eastern Area directed her to *Lampsis* while *Escanaba* continued toward *Monte Palomares*.

Captain Dimitrious reported to *Androscoggin* that his ship had cracks in the main deck and flooding in the no. 3 hold, which held a cargo of manganese ore. The no. 3 hold was divided from the boiler room (where steam was made to provide power to the ship) by a single bulkhead. This bulkhead was flexing as the 35-foot seas caused the slurry of water and ore to surge around in the hold. Captain Dimitrious feared the bulkhead would let go, flooding the boiler room, depriving the ship of power (and thus the pumps needed to empty the water from the hold), and thereby sinking his ship.

Captain Dimitrious apparently knew his ship well, however, and believed that even if the bulkhead let go, there would still be enough time to make proper preparations to abandon ship. Thus he further reported that he would continue to attempt dewatering of the hold.

By the time *Androscoggin* arrived on scene at 2145, Captain Dimitrious' pumping efforts had come to a halt. The pumps in the hold became clogged with the heavy water–ore slurry and simply could not move the slurry over the side. *Lampsis* had no portable pumps onboard. Undaunted, Captain Dimitrious had his crew drill three small holes in the bulkhead in the hope that the water, but not the slurry, would drain into the boiler room where the boiler room pumps could then put it over the side. Meanwhile, the owners of *Lampsis* had arranged for a salvage tug, the *Albatross*, which did have portable pumps, to proceed to the ship. *Albatross*, however, would not arrive for two more days.

By the time the sun came up on the morning of January 11, the weather had started to moderate; the winds were less than twenty knots, and the seas had diminished to fifteen feet. *Androscoggin*'s captain, Capt. Emerson Hayes, Jr., launched one of his small boats and sent a damage control party to evaluate *Lampsis*' condition and see if they could help.

The cutter's damage control party found eight feet of water in the no. 3 hold and three cracks in the main deck. One crack extended from the aft port corner of the no. 3 hatch to the outboard edge of the deck on the port side, and another crack extended from the aft starboard corner of the hatch to the outboard edge of the deck on the starboard side. In other words, the main deck was almost completely split at the aft end of the no. 3 hatch. The main deck aft of the no. 4 hatch had a crack that ran completely across the ship, and the boat deck had a crack that ran completely across the ship. Though no cracks were seen in the hull, Captain Dimitrious told *Androscoggin*'s damage control party that there was at least one crack in one of the ship's double-bottom tanks.

Certainly, the weather in the preceding few days had been bad, but as Captain Dimitrious told the Coast Guardsmen, he had been through worse weather in the past without serious damage to the ship. Even so, the breakup of *Lampsis* was as much a function of history as it was of the weather.

Lampsis was a Liberty ship.

* * *

ORGANIZED AND MANAGED by the U.S. Maritime Commission, the World War II–era U.S. shipbuilding programs resulted in the construction by American shipyards of an astounding 50 million deadweight tons of cargo ships and tankers from 1939 through 1945. Among the types built were the Liberty ship and the T2 tanker. A total of 4,694 merchant-type vessels were built; of these, the majority was Liberty ships (2,708 ships; cargo, with design modifications as tanker, ore carrier, and troop transport).[1]

The construction of this fleet of merchant ships was a necessity of war, and the feat was achieved through use of standardized ship designs; new and well laid out "vertical shipyards"; a dedicated workforce of 640,000; a vigorous industrial base that provided steel plate and machinery (e.g., engines, reduction gears); an infrastructure capable of transporting such plate and machinery directly from factories to shipyards (i.e., railroads); and new methods applied in shipyards, which included welding rather than riveting of ships and a sequential build process of plate fabrication (i.e., cutting, shaping, drilling), assembly of units, erection of units on the ways, and outfitting off the ways.

Though this civilian mobilization was a remarkable accomplishment, it was not without crisis—serious structural problems began to appear in the Liberty ships in 1942 and in T2 tankers in 1943.

On November 11, 1942, cracks appeared in the shell plating of a Liberty ship, the *Jeremiah Wadsworth*, while it was at sea. Between mid-November 1942 and mid-January 1943, nine other Liberty ships developed severe fractures.[2]

In the first three months of 1943, fracture of steel plates in Liberty ships and T2 tankers began to show itself publicly, and in these cases spectacularly. On January 16, while lying calmly alongside an outfitting dock at Swan Island, Oregon, the new T2 tanker *Schenectady* suddenly split amidships (with a "bang" heard one mile away) across the deck and down each side of the hull and "jack-knifed" upward; on February 17, the Liberty ship *Henry Wynkoop* cracked across the deck and down the port-side hull while taking on cargo in New York; and on March 29, the T2 tanker *Esso Manhattan* split amidships with a "bang" and jack-knifed in a fashion similar to *Schenectady* while under way outside the entrance to New York Harbor.[3] (The crew, believing the tanker had hit a mine, abandoned ship and was rescued by the *Active*-class cutter *Kimball.*[4]) Other Liberty ships developed severe fractures throughout 1943, and a new rash of Liberty ship fracturing occurred during winter 1943–1944.[5]

It was obvious that a fracture, or crack, will stop at the edge of a riveted plate, whereas in a welded ship, a crack will run across a plate and through a weld and into the adjacent plate and continue to run across that plate through the next weld and into the next plate, and so on. Yet the cause(s) of fracturing in the Liberty ships and T2 tankers was a mystery, and the practice of welding of ships became suspect in some quarters. An initial assessment of "locked-in stresses" in the welded hulls as the cause of fracturing was later rejected after more careful analysis.

One cause was discovered by British metallurgist Dr. Constance Tipper of the Department of Engineering at Cambridge University. Tipper suspected that temperature played a role in the fracturing and studied the fractures in steel plates from affected ships. She determined that the composition of the alloy caused the steel plates used in the Liberty ships to change from ductile to brittle behavior at cold temperatures. Because of this change, notches in the edges of steel plate—where stresses collect—became susceptible to fracture. This effect was termed "notch brittleness" (or "notch sensitivity").[6]

An American investigational board discovered additional causes, which included "structural discontinuities" such as square rather than radiused corners of hatch coamings, where stresses concentrated and initiated fracture, and defective welds, which became focal points of initiation of fracture.

Various remedial measures were taken at the time, some of which were reinforcement of the square corners of hatch coamings and the riveting onto ships of new structural members called crack arrestors.

A final report issued by the Department of the Navy on July 15, 1946, provided a tally of casualties: Of the total 4,694 merchant ships built, 970 ships had suffered a structural failure, some ships on more than one occasion after remediation or repair, for an adjusted total of 1,442 ships affected. However, failure such as that described for the *Schenectady* and the *Henry Wynkoop* was not common. Of the 4,694 merchant ships, twenty-five ships experienced a complete fracture of the main deck or hull; of these, eight ships were lost at sea, and four other ships were salvaged and returned to service. The total loss of life due to fracturing was twenty-six crewmen.[7]

When international commerce picked up after the war, there was a demand for the idled Liberty ships because they were relatively modern, they were designed to carry cargo, and they were available in large numbers. Liberty ships along with T2 tankers filled a need in the world's maritime industry for decades after the end of World War II. In time, however, a few more names of ships and men would need to be added to the casualty tallies in the Navy's July 1946 report.

* * *

LAMPSIS WAS NOT an anomaly: After two decades of plying the world's oceans, *Lampsis* (launched in 1943 as the *J.D. Ross*) was breaking apart in poor but not atrocious weather because of deficiencies that had been built into the ship two decades earlier. Nevertheless, the crewmen of *Androscoggin*'s damage control party were aboard to help keep *Lampsis* afloat. They knew from what they had heard from Captain Dimitrious and from their own observations that typical pumps were not going to get the water out of the no. 3 hold. The moving parts in a typical pump, whether reciprocating or impeller, were going to clog up with the ore slurry in short order.

However, *Androscoggin*'s damage control party had another alternative, a device known as an eductor. This simple device works by the Venturi effect: By passing a high-speed, high-volume flow of liquid next to a smaller opening, any liquid adjacent to the small opening will be "sucked up" and carried away.

The eductor's size and shape is dictated by its method of operation and function. It is generally a long, heavy, narrow funnel with two opposing pipe connections at the narrow end. A fire hose is attached to the connection on one side and a discharge hose to the connection on the other side. When the fire-hose pump is turned on, the initial flow of water begins and passes through the discharge hose, and, when the wide end of the eductor is immersed in a large volume of water (such as a flooded hold), the water is sucked up from the wide end of the funnel and forced out through the discharge hose.

The eductor is excellent at moving large volumes of water. It is not suitable for routine use in places such as cargo holds because inevitably there is a great deal of water splashing about from the nature of its operation. Furthermore, the eductor will never get a space completely dry.

The big advantage to the eductor, however, is that it has no moving parts. Even solid objects that are small enough to fit through the eductor's passages can be carried out along with the discharged water. The eductor seemed to be the perfect solution for *Lampsis*' situation.

Androscoggin's crewmen returned to the cutter to pick up some eductors, hoses, and pumps to charge the fire hoses and took everything back to the Liberty ship. The damage control party connected it all, fired up the pumps, and started discharging the ore slurry overboard.

For about an hour. Then the slurry managed to jam up even the eductors. *Androscoggin*'s damage control party fought this problem for a while longer but finally gave up because more time was being spent on cleaning out the clogged eductors than on pumping out the slurry.

By 1432 on January 11, all the eductors, pumps, and hoses along with the damage control party were back aboard *Androscoggin*. The cutter's boat returned to the stricken ship one more time to transport radio repair parts and a technician to work on *Lampsis*' damaged radios. Two hours later, the boat and crewmen returned to *Androscoggin* having gotten *Lampsis* some limited radio communications. There was nothing else the crews from the cutter or the Liberty ship could do. They all just waited for *Albatross* to show up and tow the ship to safety.

Captain Dimitrious was prudent. He had his crewmen make all preparations to abandon ship even though he planned to have them all stay aboard until *Albatross* arrived. He knew *Lampsis'* condition was fragile, and though he did not want to lose her, he was wise enough to put the safety of his own crew ahead of the salvage of the ship.

Captain Hayes also knew that *Lampsis'* condition was fragile. He radioed Commander Eastern Area and told him that if the weather did not hold, the chance of *Albatross* or anyone salvaging *Lampsis* was "poor" because of the extensive cracking in the main deck and double-bottom tank and the large quantity of water sloshing about in the no. 3 hold. *Albatross* was not due to arrive for another thirty-six hours.

That night, the weather held as the winds remained out of the northwest at less than fifteen knots, though a large, ten- to twenty-foot swell was left over from the previous day's storm. *Androscoggin* was able to drift most of the time and still remain close to *Lampsis*, only occasionally getting under way to maintain a reasonable distance from the stricken Liberty ship.

THE MIDWATCH RUNS from midnight until 0400. It is perhaps the watch most hostile to human beings. Sleep is disrupted, and one spends the entire time wandering around in the dark. The watch-standers get out of bed in the dark when they go on watch, and it is still dark when they go off watch and go back to bed. There are no meals or drills or musters or inspections during the midwatch. No reveille, no taps, no evening reports, nothing breaks up the simple watch routine. The midwatch can be and oftentimes is very quiet: The rest of the crew is asleep. If the ship is not going anywhere, as *Androscoggin* was not in the wee hours of January 12, the midwatch can simply be boring.

At 0300 on January 12, the boredom broke. *Lampsis'* radioman, using the radio that had been repaired by *Androscoggin*'s technician, began broadcasting an SOS. At the same time, the signal light from *Lampsis'* bridge began flashing an SOS.

Water had suddenly begun to rush into the Liberty ship's engine room. Engine rooms, particularly engine rooms on merchant ships, are best described as cavernous. There are no intermediate decks or bulkheads in an engine room because the size of the machinery demands a large space. Consequently, only an immense influx of water will be noticeable. In thirty minutes, *Lampsis*' engine room had gone from dry deck plates to having fifteen feet of water. Nobody knew if another crack had appeared or if some piece of machinery had let go, but everybody knew that *Lampsis* was taking on a lot of water and taking it on fast. It was dark, the ship was in the North Atlantic, it had known cracks with more appearing, and now there was inexplicable flooding in the engine room. Captain Dimitrious decided it was time to get off.

Captain Hayes rousted out his crew and maneuvered *Androscoggin* close to *Lampsis*. By 0315, the cutter was less than 200 feet away from the Liberty ship. Ten minutes later, one of the cutter's motorboats was lowered away and headed for *Lampsis*. Two minutes after that, *Androscoggin*'s crew launched a raft and tied it alongside to port to make it easier to embark survivors.

The motorboat's coxswain was Boatswain's Mate Second Class Roger C. Boggs. Undaunted by the swell, the darkness, and the fact that he was near a ship that was literally breaking up, Boggs maneuvered his boat alongside *Lampsis*. Boggs encountered some problems of language barriers and some anxiety on the part of *Lampsis*' crewmen; nevertheless, he managed to get nine of her crew safely aboard.

After Boggs got those men delivered to *Androscoggin*, he went back to the Liberty ship and took off another nine. Then he did it again picking up eight. He went back once more and got the final four crewmen. In less than an hour, Boggs had made four round-trips, picking up and discharging all thirty crewmen, and none were injured in the process. For his courage and skill, demonstrated not just by his boat handling but also by his speed and efficiency in completing the rescue, he would be awarded the Coast Guard Medal.

BY 0452, *ANDROSCOGGIN*'S crewmen had stood down from their rescue stations, set the normal watch, and had begun drifting again, keeping an eye on the sinking ship.

Daylight did nothing to encourage Captain Dimitrious to return to his ship. He knew, better than almost everyone else, how badly damaged the ship was. He had seen the cracks, watched the water rise first in the hold and then in the engine room, and met with the frustration of trying to pump the water out. But it was obvious to everyone that *Lampsis* was continuing to sink, albeit slowly.

Androscoggin stayed close by all day. Even though no one on the cutter or among *Lampsis'* crew had any intention of trying to go back aboard to engage in futile salvage efforts, they could not simply leave the ship. There was a remote possibility she would stop sinking. If that happened, then *Albatross*, with her specialized salvage equipment, might be able to save her, in which case the crew, or part of the crew of *Lampsis*, would almost certainly get back aboard for the tow to port. If, on the other hand, she did sink, then someone needed to confirm that so the owners could be certain of their ship's fate.

Eventually, the flooding in the engine room shorted out the electrical panels and stopped the generators. Without power, there would be no lights on *Lampsis* when darkness fell. That added another imperative to staying with her. A large, unlit, drifting hulk in the middle of the ocean was a threat. Any ship moving through the area could collide with the unseen wreck, thereby sending both to the bottom.

By sunset at 1733, *Lampsis* was very low in the water. After dark, *Androscoggin*'s crew continued to keep track of the Liberty ship by radar. At 2220, *Androscoggin* made voice radio contact with *Albatross*, and the cutter's radiobeacon was activated to enable the tug to home to the scene. Within an hour, the radar contact that was *Lampsis* disappeared. *Androscoggin*'s crew turned on the cutter's 24-inch-diameter searchlights and closed in on the position where the contact had last been. There was nothing. Rain squalls limited visibility, but it seemed the Liberty ship was gone.

There was no spectacular conclusion. No stern in the air with a propeller flailing helplessly; no bow pointed skyward as air rushed out of a few unflooded compartments. No dramatic death roll. The ship just sank.

The next morning, January 13, *Albatross* and *Androscoggin* found a large oil slick. That was the last trace of *Lampsis*.

Except for her thirty crewmen, who were safe and sound on the cutter. The next day, January 14, *Androscoggin* rendezvoused with *Barataria*, which was returning from the *Monte Palomares* search area.

All thirty men were transferred from *Androscoggin* to *Barataria*, and *Barataria* continued on to offload the *Lampsis* crew and return to the Bermuda SAR Standby.

Androscoggin proceeded to Ocean Station Echo.

THE FINAL JOURNEY of the *Smith Voyager* would sound like a comedy had it not ended with tragedy. Before leaving port, she ran aground, she had a list from poorly stowed cargo, her evaporator (which provides freshwater for drinking, washing, and cleaning) was not operating properly, she carried about 30 percent of her maximum fuel, water leaked in around the rudder, one of the generators was partially disassembled, and doors and portholes leaked. Yet *Smith Voyager* was not a rust-bucket tramping around the globe's backwaters. She was a U.S.-owned, -operated, -crewed, and Coast Guard–inspected ship of 7,600 tons.

On December 2, 1964, the crew of *Smith Voyager* began to make preparations to take on a cargo of wheat in Houston for carriage to Calcutta. As was the practice then, stevedores came aboard to install "feeders" and bins and shifting boards, which were designed to help load the bulk cargo evenly and then keep it in place once the ship sailed.

After a week, everything was installed, and *Smith Voyager* got under way to shift to a berth at a grain elevator for loading of the wheat. The ship had a pilot aboard and two tugs assisting, one at the stern and one at the bow. Apparently, there was a misunderstanding between the crew and the pilot about an engine order. *Smith Voyager* grounded gently and stayed there for about an hour until the tugs managed to refloat her.

During that same week, the ship suffered a small fire in the outboard generator. The inboard generator was started, but a few hours later it stopped running properly when the governor that controlled its speed failed. This generator had to be controlled manually until the outboard generator was cleaned up, repaired, and placed back in operation. Meanwhile, the ship suffered several blackouts as electrically operated winches overloaded the single inboard generator causing circuit breakers to trip. Consequently, when *Smith Voyager* sailed from Houston, the inboard generator was inoperative because the governor was disassembled for repair.

Cargo loading was completed early in the morning of December 12. At least two people, one of the ship's officers and an independent surveyor (checking to ensure the accuracy of the weight of the cargo), found the ship had a slight (one- to two-degree) list to starboard. Nevertheless, she was on her marks (i.e., loaded within limits though loaded right up to those limits) and sailed for Calcutta, India, at 0555.

Smith Voyager was scheduled to take on fuel in Freeport, the Bahamas. She left Houston with only 1,500 barrels of fuel (keeping in mind that she was, at the time, loaded to her marks) and then took on 4,000 barrels of fuel in Freeport. As the Coast Guard Board of Investigation would later conclude, this meant *Smith Voyager* was overloaded. And when she sailed from Freeport, other members of the crew began to notice the slight list to starboard.

After *Smith Voyager* left Freeport on December 15, her crew noticed a continuous leak around the rudder stock. The crew tightened up the gland around the stock but to no avail; on the morning of December 16, the steering gear room was flooded. There was no installed pump in the space, so the crew started a bucket brigade to bail it out. This was obviously a slow and laborious project, so the first assistant engineer cut two holes in the bulkhead between the steering gear room and the next space forward to drain the water. He planned to go back later and seal the two holes. He did not.

Smith Voyager sailed on, eastbound. By 0700 on December 20, she had a list to starboard of two to five degrees. Captain Frederick W. Mohle attributed this to the strong north-northeast wind on the ship's port side.

AS THE WIND and seas quickly built, the ship began to roll moderately and then occasionally more heavily. By midmorning, many of the crewmen noticed that the ship was listing to starboard, but they either disregarded it because the ship was rolling twenty degrees to starboard or they shared the captain's opinion that it was caused by the weather.

Captain Mohle, however, was beginning to doubt his earlier opinion. Shortly before noon, he asked the engineers if they had shifted any fuel or water because he now believed the ship had a ten-degree list. The engineers reported having shifted nothing.

Smith Voyager was powered by steam. Just about noon, the engineering crew found a steam leak in the main steam line. The only way to repair this was to shut down the steam going through the line, which meant shutting down the propulsion. *Smith Voyager* would have to be stopped.

Then, at 1230, she took a big roll. Worse, she did not come back. The chief mate and the boatswain checked the first three holds to see if the cargo had shifted. Everything appeared normal. They could not check the two holds aft because the escape trunks in them had been filled with wheat. *Smith Voyager*, powerless and beam to the wind and seas, continued to roll, and the ship's list now increased to thirty degrees.

Shortly after 1300, Captain Mohle directed the radio officer to contact other vessels in the area and ask them to stand by for a possible SOS.

Meanwhile, in the engine room, one of the watchstanders noticed smoke near an electrical distribution panel and, when he looked behind it, saw water spraying in from what seemed to be a crack in the hull.

Shortly thereafter, the radio officer lost his main electrical power. Captain Mohle appeared and told him that the no. 4 hold had opened up and instructed him to send an immediate call for help. Using emergency power, he did so.

Now came the first piece of good news for *Smith Voyager*. The motor vessel *Mathilde Bolten* was less than eight miles away and had the *Smith Voyager* on radar: She was on her way. *Smith Voyager*'s crew, however, would not wait.

Although the weather was adverse with forty to forty-five knots of wind and seas running up to twenty feet, the air temperature was in the sixties and the water temperature in the seventies Fahrenheit; in other words, survivable conditions. Captain Mohle ordered the chief officer to clear away the starboard boat (the port one was useless because of the list) and then to release it. All of the ship's crew except for the captain, the third mate, and two able-bodied seamen (i.e., thirty-eight of forty-two crewmen) abandoned ship into the oar-powered boat.

It was a disaster. The boat took on huge amounts of water, the crew could not row it, men were swept in and out, and the boat barely avoided being smashed against the side of its own ship. Somehow, the crew managed to get the boat away from *Smith Voyager*. Soon, *Mathilde Bolten* arrived, and her captain brought his ship carefully alongside the swamped lifeboat. As *Smith Voyager*'s crewmen began to stand in the swamped lifeboat to scramble aboard the rescuing ship, the boat capsized and began to crush men against the side of *Mathilde Bolten*. Four of *Mathilde Bolten*'s crew—Gabriel Ciesillski, Eckart Linke, Alber Suhr, and Horst Wolfe—went over the side, into the boat, and even into the water to help the helpless. They held the victims' heads out of the water and attached lines to them so they could be hauled on deck. It was a heroic effort, but all could not be helped. By the time everyone was aboard, four men had died.

MEANWHILE, the *Casco*-class cutter *Rockaway*, manning Ocean Station Echo, was directed to proceed to *Smith Voyager*'s reported position. The Coast Guard also called the *Secretary*-class cutter *Spencer* out of the Bermuda SAR Standby to assist. Furthermore, the *Hoegh-Fulmar*, another nearby merchant ship, diverted to help her fellow mariners.

Once the *Hoegh-Fulmar* arrived on scene, her captain agreed to stand by the *Smith Voyager* until one of the Coast Guard cutters could get there, thus permitting *Mathilde Bolten* to depart at 2130 bound for San Juan, Puerto Rico.

Rockaway arrived on scene at 0900 on December 21 and began circling *Smith Voyager* to assess the situation. Using flag signals and the cutter's bullhorn, Cdr. Marcus. H. McGarity, *Rockaway*'s captain, confirmed that there were still four men on the stricken freighter: the captain, third mate, and the two seamen. With thirty-four survivors and four bodies on *Mathilde Bolten*, this meant everyone was accounted for; there would be no need to begin a search for missing men.

Still remaining, however, was the problem of how to get the four men aboard *Smith Voyager* off their ship and onto the cutter. The grain ship's list meant that the starboard-side main deck, which was on the leeward (downwind) side, was constantly washed by the twenty-foot seas. Furthermore, there was a lot of flotsam from *Smith Voyager* floating down from the starboard side including various lines, boat falls, Jacob's (rope) ladders, two drifting rafts, and another single raft moored alongside. Whereas cutters had in the past been able to get close aboard a sinking, listing ship to pass a line and haul the survivors over in a raft, all of *Smith Voyager*'s debris made such an operation extremely hazardous. If *Rockaway* caught even one of those lines in her screws or rudder, she would become helpless while being effectively tethered to a sinking ship.

By early afternoon, *Rockaway*'s crew had recovered the two drifting rafts thus solving part of the problem. By late afternoon, the overcast seemed to hasten darkness as sunset was just after 1700.

Captain Mohle, his mate, and two deckhands, having watched the rest of the crew abandon ship twenty-four hours earlier with the deaths of four men and now watching the waves wash over their ship's main deck, knowing that water was pouring in through the non-watertight doors and portholes to starboard as well as through the no. 4 hatch, decided it was time to leave. At 1640, they fired a red flare and hoisted flags indicating their intention to abandon ship. Now.

Eighteen minutes later, *Rockaway* launched a pulling boat with a crew of twelve aboard. This may seem like a large number of men, but it was going to take a lot of manpower, not to mention skill, to maneuver that boat through the forty-knot winds and twenty-foot seas.

Lieutenant John J. Cadigan was the officer in charge of the boat, and he had his work cut out for him. He had to thread his way through the debris without fouling the boat in any of it, get alongside *Smith Voyager* without being crushed by the seas breaking across the ship's deck, pluck the four men off, and then return to his own ship.

He wisely decided that taking the boat alongside the grain ship was imprudent. Instead, he maneuvered to within twenty-five feet of the vessel and directed *Smith Voyager*'s crew to jump into the water (which had a temperature of 70°F) and grasp lines that the boat crew had deployed for them. Then he had his crew pull the boat away from the ship and the debris until they had dragged the survivors more than one hundred feet away from all the danger. They then hoisted the men into the boat and commenced the arduous pull back to *Rockaway.*

Somewhere in all this jumping and dragging and pulling, Captain Mohle fractured three ribs. But he was alive as were the other three members of his crew. Ninety minutes after *Rockaway's* crew launched the pulling boat, it was back aboard with all of its boat crew and the four survivors.

The crew of *Rockaway's* pulling boat maneuvers off *Smith Voyager.*

WHAT IS PARTICULARLY striking about the effort of *Rockaway*'s boat crew is the contrast it presents with the *Smith Voyager* crew's handling of their own boat. The conditions were nearly the same because they were only a day apart: winds of forty knots, fifteen- to twenty-foot seas, and both crews using pulling boats.

The survivors' descriptions paint a picture of disorder aboard *Smith Voyager*'s boat both during and after its launch. No one checked to see if the boat plug—typically removed to let rainwater drain out of the boat while it is stored on deck—had been put in the boat's bottom before launch. A missing boat plug would allow water to come in through the boat's bilges.

The first few men boarded *Smith Voyager*'s boat using the lifelines attached to the davits. Then, as the boat drifted back alongside the ship, other crewmen moved aft and lowered themselves into the boat from there.

Three men simply jumped from the deck into the ocean. Two of their shipmates, Able-Bodied Seaman Donald C. Covert and Boatswain James A. Wright, leaped out of the boat, swam to their aid, and pulled them back toward the survival craft where they were all finally hauled aboard.

As seas broke over the boat, it began to fill with water, and the men in it were unable to pump or bail it out. Some crewmen found it difficult if not impossible to row and simply let their oars float away, leaving the boat with limited propulsion. Some of the ship's officers struggled with the remaining oars and managed to move the boat away from *Smith Voyager*'s stern. When the boat finally drifted clear of the ship, it broached (i.e., rolled heavily) and filled almost completely with water.

The crew's efforts were clearly uncoordinated: *Mathilde Bolten* had to maneuver to get alongside *Smith Voyager*'s boat where the pulling boat should have approached the larger, ungainly ship. After *Mathilde Bolten* got alongside the swamped craft, a few of *Smith Voyager*'s crewmen leaped out of the boat grabbing lines and nets that were slung over *Mathilde Bolten*'s side. Then several crewmen all moved to the boat's side simultaneously, causing it to capsize. A few men managed to right the boat. But as all this leaping and scrambling and capsizing and righting was going on, men were being crushed or drowned.

There was understandable panic among some of *Smith Voyager*'s crew: Their own ship was sinking, the weather was poor, and they were not used to being in a boat, especially a pulling boat, in the middle of the ocean. Although some of the officers attempted to organize the survivors and get them aboard *Mathilde Bolten*, their efforts were manifestly ineffective. Had the *Mathilde Bolten*'s four crewmen not gone in the water to help *Smith Voyager*'s crew, many more might have died.

Rockaway's pulling boat, on the other hand, was launched and recovered without drama. The boat worked its way through the same seas with all the oars synchronized and the boat under complete control. Lieutenant Cadigan's crew remained calm and collected. They never seemed to consider the fact that they were risking their own lives to save strangers.

A SALVAGE TUG, the *Marinia*, had been dispatched by *Smith Voyager*'s owners to recover their ship even as *Rockaway* was making preparations to recover the survivors.

Smith Voyager's captain and three crewmen remained aboard *Rockaway* until December 23. That day, the cutter *Spencer* relieved *Rockaway* of those rescuees and of her duties as on-scene commander.

Rockaway returned to Ocean Station Echo.

As the weather abated on December 24, *Spencer* transferred the third mate and the two seamen back aboard *Smith Voyager* where they remained until *Marinia* arrived. *Marinia* arrived on scene on December 25 along with another tug, the *Elbe*. *Marinia* took the third mate and two seamen off *Smith Voyager* and Captain Mohle from *Spencer*, and on December 26 *Marinia* and *Elbe* took the grain ship in tow. *Spencer* then headed back to the Bermuda SAR Standby. On December 27—seven days after hurriedly being abandoned—*Smith Voyager* sank before she could reach port under the tow.

SMITH VOYAGER'S Donald C. Covert and James A. Wright each were awarded the Silver Lifesaving Medal for successfully recovering their three shipmates and getting them back into the lifeboat. The U.S. government was, however, prohibited from awarding such medals to foreign nationals. Consequently, West Germany awarded gold medals to three of *Mathilde Bolten's* crew, Gabriel Ciesillski, Alber Suhr, and Horst Wolfe, for their valiant efforts in getting *Smith Voyager's* crew out of the swamped and then capsized lifeboat. *Mathilde Bolten* crewman Eckart Linke was a U.S. citizen and thus was awarded the Silver Lifesaving Medal. For his masterful handling of *Rockaway's* boat and his demonstrated initiative, Lieutenant Cadigan was awarded the Coast Guard Medal.

ENDNOTES

1. For a history of the shipbuilding programs along with statistics, see the source Lane (1951). A 2001 reprint of the work is available.

2. Elphick (2006, p. 147).

3. Lane (1951, pp. 544–545).

4. Elphick (2006, p. 154).

5. Elphick (2006, p. 157).

6. Hayes (1998); Elphick (2006, pp. 170–171).

7. Elphick (2006, p. 145); Lane (1951, pp. 565–573). The title of the July 15, 1946, report is "Final Report of a Board of Investigation Convened by Order of The Secretary of the Navy to Inquire into The Design and Methods of Construction of Welded Steel Merchant Vessels, 15 July 1946."

7

BERMUDA SAR STANDBY

32°17' N, 64°47' W

ALONG WITH THEIR deployments to such parts of the world as Ocean Station Bravo in January, the Atlantic-based ocean station cutters were assigned search-and-rescue (SAR) standby status at St. George's, Bermuda, until the middle of the 1960s. This effectively put another cutter far at sea without the rigors of actually being under way. Because the cutter on this duty could either be moored to a buoy or alongside a pier, most of the crew could go on liberty.

There was, however, one catch: The cutter had to be able to get under way in less than two hours. Consequently, the cutters in Bermuda generally kept at least one third of the crew aboard to make sure the machinery and other equipment was ready to go on very short notice. Still, it was a challenge to get everything "lit off" and enough personnel aboard to man required stations within the time limit. There were no cell phones then, and recalling the crew consisted of sounding the cutter's whistle, hoisting a recall flag, flashing the cutter's light, and having the shore patrol tell the Coast Guardsmen to get back aboard.

The *Casco*-class cutter *Rockaway* relieved *Campbell* of SAR Standby duty early on the morning of September 20, 1958, and moored in St. George's Harbor at 0700. It was a Saturday, and, as was the norm for a cutter away from homeport on Saturday, there would be a commanding officer's inspection of the cutter. Soon after the inspection, the crew was mustered at quarters, and then two thirds of the crew was granted liberty.

At 1828 that evening, *Rockaway* received notice that a U.S. Navy flying boat, P5M no. 127711, had ditched approximately 160 miles north of Bermuda. Amazingly, less than forty-five minutes later, *Rockaway*'s crewmen aboard had gotten the engineering plant started, the electronics running, and enough of the other crewmen back aboard so that she could get under way headed for the aircraft.

Fortunately for the flying boat and her aircrew, the weather had been excellent when they had been forced to land on open water: winds were out of the north at less than ten knots, and seas were running little more than two feet. The Navy aircrew was also fortunate in that a commercial ship, the *Bjorgsund*, was nearby and able to pick them up. Nevertheless, the landing had been "hard," and the aircraft's pilot had suffered back injuries and possible broken ribs. With the Navy aircrew aboard, the captain of *Bjorgsund* agreed to stand by the aircraft until *Rockaway* could arrive.

The benign weather allowed *Rockaway* to proceed at full speed with all four engines on the line and the ship heading north at 17.5 knots. Consequently, she arrived on scene at 0545 on September 21. *Rockaway*'s crew put two small boats in the water, picked up the Navy fliers from *Bjorgsund*, and got the boats back on deck.

Because of the Navy pilot's injuries, he was sedated and allowed to rest. *Rockaway*'s captain, Cdr. Loy W.A. Renshaw, looked at the flying boat, looked at the weather, and asked the Navy co-pilot if he thought the P5M could be towed back to Bermuda. The co-pilot demurred; he believed his craft had suffered a broken keel and would not survive anything more than a very short tow.

Overhead was a Coast Guard aircraft out of Bermuda. Along with an ocean station cutter, the Coast Guard kept an aviation unit posted at Bermuda for SAR work. This unit (sometimes it was designated as an "Air Station" and other times as an "Air Detachment") included a shore-based rescue coordination center as well as a handful of aircraft.

THE AIRCRAFT CIRCLING *Rockaway* was not just any aircraft: It was another P5M of the same type as the Navy aircraft. Captain Renshaw began to converse over the radio with Cdr. Ira H. McMullen in Coast Guard P5M no. 1297. Commander McMullen believed the aircraft could be towed if a thorough examination of its structure showed it to be sound. So as *Rockaway* approached the Navy P5M, the Coast Guard cutter captain and the Coast Guard aircraft commander discussed the mechanics of how best to take a flying boat under tow by a cutter.

They followed the physician's maxim: First, do no harm. In this case, it meant some of the crew would have to inspect the Navy P5M to see if the keel really was broken or if the craft was otherwise structurally unsound. That meant someone would have to go from *Rockaway* to the flying boat. And *that* meant one of *Rockaway*'s small boats would have to go alongside the flying boat.

Large flying boats such as the P5M are designed to land on and take off from the water; therefore, they have comparatively strong "hulls" (their fuselages), which can take the impact of what amounts to a belly landing. Because their wings must stay clear of the water if they are to get airborne again, the wings are mounted on top of the fuselage. Finally, they have floats attached to the wings by means of struts so that the wings do not droop into the water after the aircraft has slowed down to land or has gained enough speed to take off.

Because they are aircraft, however, flying boats are designed to spend their time flying, which means they must have comparatively light weight. So whereas the bottom part of the hull/fuselage is comparatively strong, the rest of the aircraft structure is comparatively light.

A ship's boats, on the other hand, are designed and built to take adverse seas and the inevitable beating that comes with being lowered from and hoisted aboard ship. Therefore such boats, especially in 1958, were not lightly constructed. If a small boat approached a flying boat, damage to the aircraft by the small boat was far more likely than damage to the small boat by the aircraft.

CAPTAIN RENSHAW had his crew construct substantial padding and lash it to the gunwales (outboard, uppermost section) of *Rockaway's* no. 4 boat. He gave clear and explicit instructions to the boat crew: They were to approach the flying boat only from astern to make sure they did not hit the floats or struts while also avoiding the craft's horizontal stabilizers and tail.

Simultaneously and again based on the conversation between the two Coast Guard commanders, *Rockaway's* crew began to assemble a "towing bridle." Because the flying boat had such little draft (i.e., very little of the hull was in the water), it would tend to skid around the surface of the ocean if it was towed by a single line attached to its nose. This is rarely a problem when towing another ship or boat because vessels have a much greater portion of their mass submerged and will usually follow docilely behind the towing vessel. Consequently, the towing bridle was composed of three different lines, or legs: one attached to a wing, one attached to the nose, and one attached to the other wing. Those three legs of the bridle would then be joined together at the opposite end and attached to the hawser, which would be attached to the cutter's towing bitts on the aft deck.

Because the P5M had a fitting on the nose to take a tow line, no one was too concerned about attaching the center leg of the bridle to it. The P5M's designers, however, never anticipated towing the craft 180 miles, and so they created no structure or fittings elsewhere on the aircraft to take additional towing lines. Yet Commander McMullen explained to Captain Renshaw that the P5M's outer wing flap hinges should be strong enough to take the strain of the remaining two bridle legs. The cutter's crew just had to get wire straps around those hinges and shackle them to the two bridle legs. If everything was properly measured (both of the wing lines equally long and proportionate to the length of the bow, or nose, line), then all three legs would have equal strain and the flying boat should then follow the cutter like a dog on a leash.

Long-distance tows are usually accomplished using wire as a towing hawser and sometimes even chain for part of the towing connection. The reason is that a long-distance, heavy tow must be able to

Rockaway's crew prepares to tow the Navy P5M.

take the shock of the towing and towed vessels moving differently as they sail through the seaways. The heavy hawser inevitably has a dip, or catenary, to it when strung between the two vessels, which provides enough slack to act as a shock absorber.

If used as the hawser, wire—which is heavy and has very little stretch—could pull the flying boat underwater if not apart while it was being towed. *Rockaway*'s crewmen would have to use something lighter, and they had 600 feet of hemp line onboard, which would do nicely.

The bridle, though, would be one of the keys to success, and *Rockaway* was fortunate. Synthetic line was not in common use in 1958, but *Rockaway* had some: two 250-foot-long spools of 1½-inch nylon line. The nylon was very light in weight for its strength, and more importantly it would stretch. The nylon would act as the critical shock absorber in the towing rig. The crew used the nylon line for the towing bridle and made the nose leg 85 feet long and each wing leg 125 feet long.

The padded no. 4 boat was launched from *Rockaway* at 0813, and the investigating crew proceeded to the aircraft. Aboard the cutter, work continued on the bridle and the hawser itself. At 1000, the investigators reported the flying boat had taken on no water and that it appeared structurally sound.

Rockaway's crew carefully and methodically first loaded the 600 feet of hemp hawser and then the towing bridle into the no. 3 boat and launched the boat. The crew in the no. 3 boat passed one leg of the bridle at a time to the crew of the padded no. 4 boat, who then approached the flying boat and passed those lines to the men aboard, who shackled each leg in place.

Everything went flawlessly. Within an hour, all three legs of the bridle were attached to the flying boat at one end and to the hawser at the other. The crew of the no. 3 boat backed away from the craft paying out hawser as it worked its way back toward *Rockaway*. Once the no. 3 boat got close enough, *Rockaway*'s crew passed another hawser to the boat, and the boat crew spliced the two lines to make an appropriately long tow line. The no. 3 and no. 4 boats were hoisted aboard, and within another hour *Rockaway* was under way with a flying boat in tow, bound for Bermuda, 180 miles to the south.

ALTHOUGH THE CUTTER had steamed north at full speed, *Rockaway* was not going to be able to head south to return to Bermuda at a similar pace. Instead, only one engine on one shaft was engaged for minimum speed, just over four knots. This was an operation that required finesse and care. The flying boat was a delicate craft, if not a fragile one. The wing hinge mounting points might have been strong, but they were not designed for pulling the aircraft. *Rockaway*'s crew had never towed a flying boat before, so they were learning as they went along.

As the afternoon turned to night, the wind continued to blow from the northwest at less than ten knots, and the seas remained very low. As night fell, however, the wind increased, and *Rockaway*'s crew learned the first new lesson: This aircraft could not be towed downwind at the increased wind velocities. Despite the bridle, the flying boat veered badly.

Eventually, *Rockaway*'s crew figured out that the high tail of the flying boat combined with the fact that the forward part of the fuselage was immersed made the aircraft try to head up into the wind, sometimes referred to as "weather-cocking." As the flying boat veered to starboard, with the wind on its starboard side, the port wing would dip thereby immersing the port float and potentially capsizing the aircraft or tearing off the float and strut.

Rockaway altered course to starboard to 210 degrees true while the wind backed from northwest to west-southwest, or about 250 degrees true. The flying boat rode better but it continued to try to head directly into the wind.

By 0700 on the morning of September 22, the wind had increased to fifteen knots out of the south-southwest, and Captain Renshaw believed he could do nothing but head into the wind. The flying boat was still safe and rode well enough this way, but they were no longer headed toward Bermuda, which was almost due south. *Rockaway* was in a quandary: The captain did not want to cast off the tow and probably lose the flying boat. But the only way the cutter could keep towing without breaking the aircraft apart was by heading away from Bermuda.

Shortly before 0900, the Coast Guard P5M arrived overhead, and the pilot took a look at the situation and suggested that *Rockaway* attach a drogue to the flying boat's port float. A drogue, sometimes referred to as a sea anchor, is nothing more than a large funnel made out of cloth. A ring of metal or plastic holds the wide end of the funnel open. The wide end is attached to a line that is attached to the vessel (or, in this case, the flying boat). As the vessel moves through the water, the immersed drogue creates drag. By attaching a drogue to the port float, the flying boat would be forced to turn to port, and this might allow *Rockaway* to gain a proper course.

Captain Renshaw was willing to try this because the weather continued to pipe up, and he was concerned that if he did not get the flying boat to Bermuda soon, he might not get it there at all. *Rockaway* launched her small boat, a crew attached the drogue, and Captain Renshaw tried to come to a new course.

It did not work. The wind increased to twenty knots, the drogue collapsed, and *Rockaway* was forced back to a more westerly course.

At 2000 that night, Captain Renshaw made an astute decision: He would quit fighting the elements. Instead, he would simply let the whole rig—cutter and flying boat—drift northeasterly until he had made enough leeway so that he could resume a southwesterly course and enter St. George's.

Yet within two hours, the barometer began to go up while the wind and seas calmed down. By midnight, *Rockaway* was steering southeast directly for St. George's. At 0630 on September 23, *Rockaway*'s crew began shortening the tow as the cutter approached Bermuda. By 0700, the tow had been passed off to a small Navy tug, which took the flying boat the rest of the way in from where it would fly again. At 0922, *Rockaway* moored to a Coast Guard mooring buoy in St. George's Harbor.

At 1330 that afternoon, two thirds of *Rockaway*'s crew went on liberty.

ROCKAWAY WAS NOT the first Coast Guard cutter to tow a Navy flying boat while on Bermuda SAR Standby. The cutter *Casco* was moored to a buoy in St. George's Harbor on Friday, February 17, 1956, when, at 1340, the crew received a distress call relay from the Coast Guard Air Detachment reporting that Navy P5M no. 6507 was probably going to ditch south of the island. Forty-five minutes later, *Casco* was unmoored and under way. Forty-five minutes after that, the P5M ditched successfully. By then, *Casco* was making full speed toward the flying boat.

Sunset was at 1802. By 1955, the crew of *Casco* had the flying boat in sight: It was being looked after by two merchant ships, the Italian freighter *Versilia* and the Canadian ship *Constructor*. *Casco*'s commanding officer, Cdr. James S. Muzzy, released the two ships and launched one of his starboard-side boats along with a raft to pick up survivors.

The weather was favorable with about twenty knots of wind out of the west-southwest and very moderate seas. The boat made three trips to the flying boat and used the tested method of floating a raft from the boat to the aircraft to pick up the aircraft's people. Less than an hour after arriving on scene, the crew of *Casco* had successfully recovered sixteen people from the flying boat, and five of the aircrew remained aboard the P5M to assist with the towing evolution.

Casco's small boat carried a 6-inch manila towing hawser from the cutter to the aircraft's nose fitting where it was secured. Shortly after 2300, the small boat picked up the remaining aircrew, and *Casco* started towing the flying boat north.

On February 18, the cutter continued to tow the flying boat north at about five knots with wind speed and direction remaining favorable to the tow. Because it was Saturday, at 1000 the cutter held a commanding officer's inspection. Early on the morning of February 19, *Casco* began maneuvering to turn the tow over to a small Navy tug just off Bermuda. Shortly after 0900, the tow was shifted, the aircraft personnel transferred to smaller Navy boats, and *Casco* headed home to Boston.

NOR WAS *CASCO* the first Coast Guard cutter to tow a Navy flying boat while on SAR Standby. On May 29, 1952, the cutter *Androscoggin* was on two-hour SAR Standby in Miami taking on bread, milk, and meat when, at 1430, the crew was notified of a downed Navy PBM near Orange Cay in the Bahamas. Thirty-two minutes later, *Androscoggin* was under way. Within an hour, the cutter was at full speed heading for the flying boat.

By 1730, *Androscoggin* arrived at the aircraft's location and began taking the passengers and crew off. Within ninety minutes, everyone from the PBM was recovered and aboard *Androscoggin*, the tow line was connected, and the flying boat was headed for Florida at the end of a hawser.

The weather stayed favorable throughout the night with winds out of the southwest at about twenty knots, so the PBM followed docilely astern. Just before 0600 the next morning, *Androscoggin* passed the tow to a Coast Guard 83-foot patrol boat and transferred the unexpected passengers to a Coast Guard 38-foot patrol boat. At 0841, *Androscoggin* moored in Miami, mission accomplished.

At 1030, *Androscoggin*'s crew took aboard the rest of the groceries.

THE BERMUDA SAR STANDBY was not the only location from which an ocean station cutter could or would be dispatched on an SAR case. *Androscoggin* had been in Miami in May 1952 when her crew sailed for the aircraft tow. Just over three months earlier, the *Casco*-class cutter *Yakutat* was bound from her homeport of Portland, Maine, to Newport, Rhode Island, when the weather began to deteriorate significantly.

Early in the morning of February 18, 1952, the wind backed from northeast to north and built from forty knots to sixty knots with seas building to more than fifty feet. Attempting to sail south from Maine to Rhode Island meant *Yakutat* was heading downwind and down-sea. This was nearly disastrous as the cutter rolled heavily from side to side, and rudder control diminished. At 0332, *Yakutat*'s commanding officer, Cdr. Joseph W. Naab, Jr., ordered his cutter's course changed from south to west so that he could anchor in the lee of Cape Cod until things improved.

Shortly after 0900, the cutter was anchored near Provincetown at the very tip of Cape Cod. It was an excellent location to anchor, well protected from the winds even as they backed from northeast to northwest. *Yakutat* rode comfortably, secure from the ravages of the storm.

For four hours. Shortly after 1300, *Yakutat*'s crew weighed anchor and headed back out into the maelstrom because the Coast Guard had learned that a tanker had broken in two approximately 27 miles east of Pollock Rip Lightship.

CAPTAIN FREDERICK C. PAETZEL and the crew of the loaded, 10,000-ton tanker *Fort Mercer* were laboring through the maelstrom southeast of Cape Cod when a loud noise was heard at 0810, and oil was subsequently observed on the surface of the water on the starboard side of the ship in the vicinity of the no. 5 tank. Captain Paetzel radioed a message to the Coast Guard requesting assistance, and the crew of *Fort Mercer* was alerted. At 0905, the *Winds*-class cutter *Eastwind* (6,500 tons, 269 feet), an icebreaker, and the *Casco*-class cutter *Unimak*, which were approximately 120 miles south of Nantucket Lightship, were directed to proceed to assist *Fort Mercer.*[1]

At 1030, a second loud noise was heard aboard *Fort Mercer*, and the chief mate informed the captain that the no. 5 starboard tank was losing oil.

Fort Mercer was a T2 tanker. The ships of this class each had two superstructures. One superstructure was located just forward of amidships and incorporated the bridge. Berthing areas for the deck crew (captain, mates, radio officer, deckhands) were also contained in this forward deckhouse so that these individuals were close to the ship's bridge and radio room. There was another superstructure located on the stern above the machinery spaces, and the engineering berthing areas were contained in this aft deckhouse.

Hundreds of T2 tankers had been built during World War II to carry the oil and aviation fuel needed by the Allied fleets and aircraft. After the war, the T2 tankers became available as relatively modern ships to help replace the maritime industry's tanker fleet, which had been decimated by the same war. Some of the T2 tankers would stay in service for decades after the war, getting new power plants, having hull plates renewed, and, in some cases, cut in half to have entire new sections welded into the hulls to increase their carrying capacity (i.e., "jumboized").

But like the Liberty ships, the T2 tankers occasionally suffered structural failures of varying severity.[2]

AT 1050, *FORT MERCER* broadcast an urgent message requesting vessels in the vicinity to stand by. At 1104, the Coast Guard advised *Fort Mercer* that the cutters *Eastwind*, *Unimak*, and *McCulloch* were en route to assist. At 1113, the medium-endurance cutter *Acushnet* (1,720 tons, 213 feet) was dispatched from Portland, Maine, and directed to assist as well.

At 1140, a third loud noise was heard aboard *Fort Mercer*, and a crack in the starboard hull was observed extending above the waterline in the vicinity of the no. 5 tank. At 1158, *Fort Mercer* broadcast an SOS reporting that the hull was splitting.

Fort Mercer's SOS was issued none too soon. At 1205, the ship broke in two in the vicinity of the no. 5 tank, just behind the forward superstructure, leaving the captain, his three deck officers, and five other members of the deck crew on the bow section of the ship and the remaining thirty-four members of the crew on the stern section. Those on the bow were without power or propulsion. Those on the stern were without radios.

A rescue fleet was on the way in the form of *Eastwind*, *Unimak*, *McCulloch*, and *Acushnet*, and *Yakutat* would be directed to sail within the hour. In addition, the First Coast Guard District Office, which was controlling the overall case, directed Chatham Lifeboat Station, near the "elbow" of Cape Cod, to dispatch a 36-foot motor lifeboat to the scene.

At 1230, the wood-built 36-footer CG-36383, manned by Chief Boatswain's Mate Donald Bangs and his crew of three, headed to sea from Stage Harbor, Chatham, in conditions that had driven a 311-foot, 2,500-ton Coast Guard cutter to safe anchorage and had broken a 500-foot, 10,000-ton tanker in half. Getting under way in that sea state, coxswain and crew demonstrated the mettle of the old surfmen.

On the opposite side of Cape Cod, *Yakutat* would have to head north then east to get around the cape before heading south to the broken *Fort Mercer*. Captain Naab ordered full speed for his cutter as he headed northeast around the tip of Cape Cod. For five minutes. The violent pounding and near-total immersion of the cutter's bow into the seas made it impossible to keep up the pace. Twenty minutes later, *Yakutat* was forced to reduce speed yet again. By 1429, the cutter was able to alter course sufficiently to the east, rather than northeast, so that speed could once again be increased.

AT 1455, THE EXPERIMENTAL shore-based radar at the Chatham Lifeboat Station picked up two radar contacts 5.6 and 5.7 miles offshore. At 1500, visibility from shore momentarily improved. The officer-in-charge of Chatham Lifeboat Station sighted the bow section of a ship, reported the sighting to the Search and Rescue Center of the First Coast Guard District Office in Boston, and proceeded to North Beach with his beach apparatus[3] and part of his crew to prepare to render assistance if possible.

A Coast Guard aircraft (PBY no. 1242) was sent to investigate the sighting. Shortly after 1600, the aircrew sighted the same bow section along with a stern section, which two hulks were the offshore contacts picked up by the Chatham radar. The aircrew was able to identify a vessel name: *Pendleton*.

Pendleton was another T2 tanker. Like *Fort Mercer*, she was loaded with kerosene and heating oil but had arrived off Boston late in the previous evening, February 17, when the same blinding and violent snowstorm caused her master, Captain John Fitzgerald, to choose to head offshore to await more moderate weather before attempting to enter the port. He headed east-northeast on a slow bell stemming the worst of the seas and giving himself room to maneuver. By the early morning of February 18, the tanker was 10 to 20 miles off Cape Cod Light. At 0550, *Pendleton* took two consecutive, heavy lurches each accompanied by a loud, explosive sound then broke in two between the no. 7 and no. 8 cargo tanks.[4]

Pendleton's crew had no warning and did not get an SOS transmitted before the tanker broke in two. And once broken in two, there was no power to the radio room to allow her crew to do so. The captain, his three deck officers, and five other members of the deck crew were on the bow section of the ship and the remaining thirty-three members of the crew were on the stern section. Those on the bow were without power or propulsion. Those on the stern were without radios. No one in the world knew that the men aboard *Pendleton* were in peril until the Coast Guard aircrew identified the hulks around 1600, ten hours after the ship broke up.

Now the Coast Guard had a real challenge on its hands. Not one but two broken tankers. Not two, but four sets of possible survivors.

After the identification of the *Pendleton*, *Yakutat* was designated as the on-scene commander for the *Fort Mercer* and *Pendleton* rescue operations.

THE OFFICER-IN-CHARGE of Chatham Lifeboat Station returned from North Beach because the bow and stern sections of the *Pendleton* had not grounded, hence rescue operations could not be carried out with beach apparatus. Because the stern section of the *Pendleton* was drifting toward Chatham Bar, where it would be in danger of capsizing with loss of all hands aboard, Boatswain's Mate First Class Bernard C. Webber and a crew of three were directed to man the station's other 36-footer, CG-36500, and head for *Pendleton*'s stern.

Webber got under way from Old Harbor, Chatham, just before 1800 on February 18: It was dark, it was still snowing, it was still blowing hard, and his only means of navigation was a compass. As he crossed Chatham Bar, the breaking seas threw Webber's boat on its beam ends, smashing the windshield, tearing the compass loose, and knocking Webber to the deck, but Webber kept going. By seaman's eye, he sought the light of Pollock Rip Lightship to get his bearings, but by a stroke of providence, he found instead *Pendleton*'s stern section lurking in the pitch black dead ahead just before 1900.

Webber directed one of his crew to illuminate the searchlight of CG-36500. Upon seeing the searchlight, the crewmen on the stern section of the *Pendleton* spontaneously rigged a Jacob's ladder stretching from the deck to near the water's edge and began to descend the ladder. Webber reacted immediately to begin to pluck one man at a time from the ladder. As Webber made his approach, a crewman at the bottom of the ladder would jump into CG-36500—or into the raging sea if he mistimed his jump hence be recovered by the crew of the motor lifeboat.

Webber made repeated approaches to the huge, groaning hulk, plucking off one man at a time until he had twenty survivors, or a total of twenty-four people, on his motor lifeboat. This meant the boat's weight had effectively increased by almost 20 percent, but Webber kept going.

During every one of those approaches, the motor lifeboat would yaw and pitch and corkscrew as Webber fought to keep control and as the stern of the tanker heaved up and down and rolled toward and away from the boat. The extra weight made the boat even more difficult to handle in the heavy seas and driving wind. Webber continued maneuvering to get the last thirteen men off the hulk and into the boat. During this series of recoveries, one crewman mistimed his jump into the boat, landed in the water, and was swept by the seas toward the propeller of the stern section, where he managed to grab hold of a propeller blade. Webber maneuvered CG-36500 to recover the crewman, but a wave lifted the motor lifeboat by its stern and carried the boat directly toward the propeller blade and the crewman. Webber backed hard on the 90-horsepower gas engine of CG-36500, but to no effect, and the boat was thrown against the propeller blade and crushed the crewman.

Webber and crew finished recovering the crewmen from the stern of the *Pendleton* and prepared to head into shore. The time was 2000.

Webber was uncertain of his location, his boat was marginally stable with the excess number of people aboard, the compass was useless, he had no other means of navigation, the winds and seas had not abated, it was dark, and it was still snowing. As he had been heading into the seas when he left Chatham, he concluded that by heading down-sea he would be going in the right direction and could at least beach the motor lifeboat and get the survivors ashore. Webber proceeded down-sea, the crew later spotted the red flashing light of the buoy marking the turn to the entrance of Old Harbor, the motor lifeboat entered the harbor, and the thirty-two rescuees were offloaded at the fish pier.

If the crowd of people on the pier had not seen thirty-two survivors and four crew get off CG-36500, no one would have believed the story. Webber and all the members of his crew would be awarded the Gold Lifesaving Medal.

The stern section of the *Pendleton* would ultimately run aground about five miles south of Chatham Lifeboat Station.

THE CUTTER *EASTWIND* found the stern section of *Fort Mercer* at 1055 on February 19. The crew of *Eastwind* passed a line to the crewmen on the hulk and then passed a walkie-talkie along the line to establish communications. The chief engineer thus informed the crew of *Eastwind* that many of the *Fort Mercer* crew wished to be taken off the hulk.

Eastwind was an icebreaker and, though heavy and seaworthy, was simply not "handy" enough to go alongside the hulk and carry out a rescue of the *Fort Mercer* crew. Icebreakers have to be very beamy and very deep to carry out their primary mission, which is to smash through thick ice. That bulk, however, means they are slow to respond to engine orders, and their beam and draft makes them ponderous to turn.

Nevertheless, the crew of *Eastwind* began to rescue *Fort Mercer* crewmen by use of a life raft, which was drawn back and forth between the cutter and the hulk by means of a line. Three crewmen had been recovered by this method when the cutter *Acushnet* arrived on scene at 1344.

The captain of *Acushnet*, Lt. Cdr. John M. Joseph, received permission from *Eastwind*'s captain, Capt. Oliver A. Peterson, to maneuver *Acushnet* alongside the stern section of *Fort Mercer* and take crewmen off.

This was the type of work for which *Acushnet* had been built. She was a former Navy salvage ship that had been designed to get alongside broken vessels. Thus she had heavy hull construction, good maneuverability (with twin shafts and 3,460 horsepower), and a low freeboard, especially aft, so that the cutter's crewmen could reach whatever it was they were working on.

Captain Joseph put his cutter alongside the broken tanker's stern section at 1428. Although the weather had moderated from the day before (the winds had "diminished" to thirty knots and the seas were beginning to decrease as well), it was still not easy staying alongside the hulk. Within two minutes, five members of *Fort Mercer*'s crew managed to get aboard *Acushnet*. Then Captain Joseph was forced away. But he went back and was alongside again at 1450, and another thirteen crewmen scrambled to safety.

Seven crewmen including the chief engineer elected to stay aboard the stern section, which had maintained steam in one boiler and a functioning engine for propulsion throughout the ordeal, and six other crewmen also chose to remain aboard due to reasons, such as age, precluding them from making the jump onto *Acushnet*.

The owners of *Fort Mercer* arranged for commercial tugs to tow the stern section: *M. Moran* and *Foundation Josephine* took up the tow on February 20 and ultimately towed the stern section to a dry dock in New York.

MOTOR LIFEBOAT CG-36383 with Chief Boatswain's Mate Donald Bangs and crew aboard would arrive at Pollock Rip Lightship around 1610 on February 18 and be diverted from the *Fort Mercer* case to the bow of the *Pendleton*, which was drifting down on the lightship from northward.

Bangs headed CG-36383 for the broken bow and upon arrival circled the hulk twice while repeatedly sounding blasts on the horn, but the crew did not detect any signs of life aboard. A report was radioed to the lightship at 1700, and Bangs then steered a course for the stern section of *Pendleton.*

At 1725, the crew of the Pollock Rip Lightship sighted a light on the bow section of *Pendleton* and reported this to the cutter *McCulloch*, which then headed for the bow section. Arriving at 1745, the crew of *McCulloch* confirmed life aboard the hulk and radioed CG-36383 to return to the scene.

Bangs and crew were back on scene just before 1900 and sighted one man on the starboard wing of the bridge. *McCulloch*'s captain, Cdr. Theodore F. Knoll, ordered the cutter's crew to starting spreading oil to calm the surface of the raging seas, and CG-36383 approached the bow. The man on the bridge either jumped or was washed overboard: The crew of CG-36383 attempted to recover him, but he was swept away into the night. *McCulloch* and CG-36383 searched intensively for the man but to no avail.

The bow section of *Pendleton* ultimately grounded approximately 6.7 miles from Chatham Lifeboat Station, and a crewman from the salvage tug *Curb* would later search the hulk and find one body. The other crewmen from the bow section were never found.

THE CUTTER *Yakutat* arrived on scene at *Fort Mercer*'s bow section at 1829 on February 18 as CG-36500 was still heading for *Pendleton*'s stern section. As had been true at Chatham, the wind was blowing near fifty knots, the sea was rough, it was dark, and it was snowing, and Captain Naab attempted to maneuver in close to the hulk.

The bow section of *Fort Mercer* pitched, rolled, and heaved with the seas. The cutter also pitched, rolled, and heaved as it struggled against the seas and winds. The huge waves would break over both the hulk and the cutter but rarely at the same time. The looming presence of the tanker's broken bow section would provide *Yakutat* an occasional lee from the wind and then, as *Fort Mercer*'s bow section pitched down or rolled, take that very brief respite away.

Coast Guard cutter logbook entries are usually terse and factual. They contain little explanation but just simple statements of what was done and when. It is rare for a logbook to contain a characterization of the weather or the conditions. On this night, however, *Yakutat*'s log received an entry that read: "Heavy N.W. winds and seas makes approach difficult." Indeed they did. Speed was necessary to keep control over the cutter. But too much speed would hurl it into *Fort Mercer*'s bow section potentially sinking the hulk. If *Yakutat* got too close and struck the hulk, the cutter could be damaged, an extremely dangerous situation in the conditions.

Circling overhead, a Coast Guard B-17 Flying Fortress (PB-1G 77249) dropped flares to provide illumination while *Yakutat*'s crew made repeated attempts to shoot a line over to the bow section of *Fort Mercer*, but the wind repeatedly carried the shot lines away. Rather than waste his remaining shot lines at what were obviously futile efforts, at 2152 Captain Naab moved his cutter away from the lurching hulk to reconsider his circumstances. *Yakutat* stayed close enough to maintain signal-light contact with the survivors, and her crew confirmed that there were nine crewmen on the bow section.

YAKUTAT'S CREW came up with an innovative plan. They would tie a series of three rafts together on long lines interspersed with lights. The cutter would then tow this whole string toward the tanker's broken bow and attempt to float it across to the hulk.

In the meantime, *Fort Mercer*'s crewmen moved to the forward-most part of the hulk (which was the bow and fully out of the water) from the partially submerged deckhouse (which had been the forward deckhouse). At 2330, the nine crewmen began to make their way from the chart room in the bridge by climbing down a makeshift line of knotted signal flags, whence they would scramble up the sloping deck to the forecastle of the broken bow section. One of the many waves that swept the deck carried the radio officer away to his doom. The other eight crewmen made it to the forecastle.

By 0156 on February 19, the rafts were afloat, *Yakutat* was in position, and the crew of *Yakutat* started to float the rafts downwind toward the hulk, but fate delivered a blow: The line between the rafts and *Yakutat* parted. As the rafts swept past the bow section of *Fort Mercer*, three crewmen jumped into the water to get to the rafts.[5] The rafts, no longer attached to anything other than themselves, skidded off downwind at a speed no swimmer could match. Then, as *Yakutat* rounded to near the bow of the hulk, a fourth crewman jumped into the water.

In the blowing snow and the dark, the cutter's crew quickly lost sight of the four men bobbing in the frigid, heavy seas. The men were lost.

Yakutat maneuvered to recover the rafts. Captain Naab then made a prudent decision. He decided to take no other action until daylight. He had lost several shot lines, which could not be replaced. His attempt to float rafts had failed because the line parted, and four men had died. He would not risk the lives of the four remaining survivors aboard the hulk nor the lives of his own crew in further efforts when everything—the wind, the seas, the temperatures, the visibility, everything—was against him. He would wait.

JUST BEFORE SUNRISE on February 19, the wind abated briefly to less than fifty knots. The snow stopped but it remained overcast. The air temperature remained at freezing, and the water temperature was a numbing 40°F. Captain Naab decided that, with the aid of sunlight and the apparent, though inconsistent, moderation of the wind, it might be possible to get *Fort Mercer*'s remaining four crewmen off using the cutter's motorboat. *Yakutat* held reveille before sunrise, and her crew started making preparations.

Putting a small boat in the water from a ship is rarely easy. In the days before wire hoisting cables and high-speed winches, it was always risky. If a ship began to pitch and roll, it became downright dangerous.

As soon as the boat was raised off the deck, it was effectively suspended in the air by two sets of lines running through blocks (i.e., falls), one set at each end of the boat. This meant (and still means) that the suspended boat was free to swing. Wind alone would cause the boat to swing. If the ship had any motion at all, then that motion would be transferred to the boat, and the swing of the boat would increase. If the ship's motion became significant, the swinging would become violent.

There were additional lines attached to the falls (frapping lines) that could be used by crewmen on the ship to minimize the amount of outward swing of the boat. But of course these lines could do nothing to control the inward swing. Thus, the most important thing that was needed before launching a boat was to minimize the ship's motion to minimize the swinging of the boat.

This meant, first of all, slowing down as much as possible but still keeping enough speed to maintain steerageway. Second, to minimize the inboard–outboard swing of the boat, the ship would come to a course that would minimize roll. This almost always meant heading into the sea because heading with the sea (i.e., down-sea) would have an adverse effect on controlling the ship and would also aggravate the roll. Of course, heading into the sea maximized pitch, but this was acceptable if not desirable because the boats were located near amidships where the pitching motion was felt the least.

That took care of mitigating the motion of the boat as it was hoisted off the deck and started over the side. There still was another whole set of challenges in getting the boat into the water after it was swung out.

First, one must think about what will happen to the boat once it is in the water and alongside a ship. If the boat is on the windward side of the ship, the wind and waves will hold the boat against the ship. This will mean that, at the least, the boat will be hard-pressed to get away from the ship and, at worst, will be smashed against the ship repeatedly. Therefore, it makes sense to have the ship head so that the boat is in the lee of the ship. Because wind and waves generally come from the same direction, this means that by heading the ship almost directly into the seas, but with the wind just slightly on the side of the ship opposite the boat, the ship will have the best boat-launching course.

As the boat is suspended from the falls over the water, waves pass under the boat. As the boat is lowered, it will get to the point where it is alternately afloat (when the hull touches a wave crest) and suspended (when the trough of the wave passes).

But when the boat touches the water, it is suddenly subjected to all the forces of the waves. No two waves are ever the same, and thus every wave has a slightly different effect on the boat; one wave may push the boat slightly toward the ship, and the next wave may pull it slightly away from the ship. Furthermore, no two waves are of exactly the same height, and thus one wave may barely graze the boat's keel, whereas the next wave may have the boat fully afloat.

These few moments, when the boat is alternately afloat and suspended, alternately pulled away from the ship and pushed toward it, and still swinging in and out because of the inevitable rolling of the ship, are critical and dangerous. When the boat is afloat, the lines in the falls may go slack allowing them to entangle themselves; then, when the wave "drops out" from under the boat, the falls can no longer be lowered. Worse, slack lines can catch the fingers or hands or arms of unsuspecting sailors and then, when the wave drops out and the falls go taut, human appendages are severed.

The key, therefore, is to minimize the amount of time the boat is in this transition phase. The goal is to get the boat afloat and the falls disconnected as quickly as possible thus minimizing the probability of jammed falls or injured crewmen. Equally important, the boat needs to be released from the falls quickly (i.e., the falls slackened) so that the boat can get clear of the ship and avoid being pounded against the hull. On the other hand, releasing the falls too soon (before the boat is truly afloat) will cause the boat to drop into the ocean possibly damaging its hull and seriously injuring its crew. This is one case where, indeed, timing is everything.

The boat will be lowered deliberately (quickly, but not too quickly) until it just begins to become waterborne. The supervisor on deck continues to watch until the boat is low enough that it will just be fully afloat on the crest of a wave. Then, when the next crest touches the boat's keel, he gives the order "Let Fall," which, in one of those peculiarities of nautical terminology, does not mean the boat is permitted to drop but rather tells those tending the falls to slack off completely. The purpose of this move is to ensure the boat stays waterborne as the crest passes so that the boat will slide down into the succeeding trough without being "snatched up" by suddenly taut falls.

Thereafter, nearly as soon as the boat crew hears the "Let Fall" order, the boat crew must disconnect the falls so that the boat is no longer attached. If the timing is right by both the deck crew and the boat crew, the falls will keep a constant strain as the boat slides down the crest and into the trough and the falls are disconnected. This way, the falls never go slack while attached to the boat, the boat spends minimal time alongside the ship, and the boat is away successfully.

Getting the boat back aboard is, as repair manuals so often say, "the reverse of these procedures." First, the boat must be maneuvered alongside in the correct location so that there is minimal contact between the boat and the ship's hull. Then the boat must be hooked up to the falls while it is in the trough, and both sets of falls must be connected quickly. The boat must then be hoisted quickly so that, once again, the falls never go slack and the boat is brought above the crests before too many waves pass under it. Then the whole battle of swinging the boat in and putting it back in its cradle must be fought.

THIS PROCESS SOUNDS and is complex, and the evolution is dangerous even under good conditions. In February 1963, the cutter *Half Moon*, on Ocean Station Charlie, prepared to recover one of her boats, which had been doing routine work. As the boat came alongside, the rudder inexplicably unshipped (i.e., detached) from the boat. The coxswain then lost control and the boat surged forward, struck the cutter's anchor, capsized, and dumped all five crewmen in the water.

The crew of the cutter managed to recover them, but one, Boatswain's Mate Third Class John W. Williams, Jr., died from drowning. This was on a day with thirty knots of wind and moderate seas.

When the seas are not moderate but are running at twenty or thirty feet, and the wind is blowing a full gale, the whole evolution grows exponentially more difficult. Add rain- or snow-slicked decks, the inconsistent motion of any vessel in a rough seaway, and the noise of the wind that muffles commands, and quite soon the whole evolution seems impossible. This is why in so many cases on ocean stations, Coast Guard cutter captains elected to float rafts rather than launch boats to recover survivors from sinking ships or to pick up men overboard.

But when the rafts do not work because they carry away or conditions make it impossible to pass lines, then the captain has little choice but to try the boat. And that is what Captain Naab did. Sunrise was at 0635 on February 19, and the cutter began feeling its way around for a best possible launch course while staying as close as possible to *Fort Mercer*'s bow section. *Yakutat* maneuvered close enough to the tanker hulk so that instructions could be given by bullhorn to the four survivors. The four men were told that once the small boat got close and *only* when the boat crew directed, they were to jump into the water one at a time to be picked up by the boat.

After the instructions had been passed, the boat crew prepared, and the best compromise course reached, the boat-launching crew went to work at 0830. *Yakutat*'s launching evolution under these conditions was described by one *Fort Mercer* survivor as a "masterful piece of seamanship."

It was, but it had not been without incident. The boat had taken a beating while becoming waterborne and began to take on water, yet it got away successfully with a crew of five: Ens. William R. Kiely, who was the coxswain, along with Ens. Gilbert E. Carmichael, Engineman Second Class Paul Black, and Seamen Edward Mason and Webster Terwilliger.

Ensign Kiely got his boat perilously close to the tanker hulk and told the first crewman to jump. This man was Captain Paetzel, who was in poor physical condition and thus jumped first. At the moment

Yakutat's motorboat lowers away for the bow section of *Fort Mercer.*

Yakutat's motorboat crew returns with two survivors.

Captain Paetzel entered the water, one sea lifted the boat and hurled it toward the tanker hulk. It seemed as if the captain was about to be crushed against his own broken ship, yet Ensign Kiely's "superb handling" of the boat averted the killing blow, though he and his craft were still slammed hard against the tanker hulk thus springing further leaks in the wooden hull. Ensign Kiely got clear, made another approach, and finally the four boat crewmen pulled the nearly unconscious captain from the water.

Ensign Kiely called for the next man to jump, and he, too, was recovered successfully. Now, though, the boat crew realized that they were not just "taking on some water" from above and below. The motorboat was truly sinking and could only stay afloat for a few more minutes. It would do no good to get the remaining two survivors on the boat if, after that, all nine had the boat sink from under them. Ensign Kiely headed back to his cutter.

He got alongside, he got hooked up, and he got hoisted. The boat was nearly destroyed, but the men were all alive. For Ensign Kiely's "considerable daring and skill," he was awarded the Gold Lifesaving Medal, and each member of his boat crew was awarded the Silver Lifesaving Medal.

BUT THERE WERE still two men on *Fort Mercer*'s bow section, a floating hulk that showed increasing instability. The weather that had seemed to moderate around dawn now deteriorated again as the wind built to more than fifty knots. *Yakutat*'s motorboat had been all but destroyed in a period of less than one hour in these conditions. Launching another boat seemed unnecessarily dangerous, even foolhardy. Once again, Captain Naab had to make a decision, and once again his decision making was sound. He would not send another boat. He would try, one more time, to shoot a line then pass a raft to the hulk.

Shortly after 1000, *Yakutat* successfully passed a shot line to *Fort Mercer*'s bow section. By 1015, a raft was in the water being hauled toward the hulk. By 1035, both remaining survivors were in the water alongside the raft. Seven minutes later they were both in the raft, and

Yakutat maneuvered as her crew heaved around on the line. At 1051, the raft was alongside, and at that very moment *Fort Mercer*'s bow section capsized and began to settle. At 1101, the two survivors were on *Yakutat*'s deck alive and as well as could be expected.

Yakutat stayed with *Fort Mercer*'s capsized bow section to warn other ships away until shortly after midnight on February 20 when the cutter was relieved by *Unimak*. Then, with Ensign Kiely standing the watch, *Yakutat* headed for home in Portland, Maine.

Yakutat's raft being hauled to the bow section of *Fort Mercer*.

Yakutat crewmen help a *Fort Mercer* survivor aboard.

At 1650 on February 20, by authority of Commander Eastern Area, *Unimak* opened fire on the bow section of *Fort Mercer* and sank it as a menace to navigation.

ENDNOTES

1. Times and details of events with respect to the *Fort Mercer* rescue operations are largely taken from the report U.S. Coast Guard Commandant (1952a). Other sources are listed in the Bibliography.

2. See discussion in Chapter 6.

3. *Beach apparatus* was the gear, such as the Lyle gun and the breeches buoy, stowed in a *beach cart* and used by surfmen and later Coast Guardsmen to effect rescue of the shipwrecked when immediately off a coast. For a description and illustrations of beach apparatus, see "The Beach Apparatus Gear" at the U.S. Life-Saving Service History website: http://www.lifesavingservice.org/beach_apparatus_gear.html. Accessed 1 June 2009.

4. Times and details of events with respect to the *Pendleton* rescue operations are largely taken from the report U.S. Coast Guard Commandant (1952b). Other sources are listed in the Bibliography.

5. The report U.S. Coast Guard Commandant (1952a) describes the rafts episode, but the sequence of events is not presented. Frump (2008), based on public comments of Captain Naab, clarifies the sequence of events: the line parted; the rafts, taken by the wind, swept past the bow section; the three crewmen jumped for the rafts.

8

NOVEMBER

30°00' N, 140°00' W

THE COAST GUARD'S VIEW of early aircraft as "flying lifeboats" was the reason the service became involved with aviation and also why the Coast Guard's fleet of aircraft consisted mostly of flying boats until World War II. Not willing to argue with success, the service continued to use flying boats into the early 1980s and had helicopters capable of water landings until 1994.

The Coast Guard adopted this technology because it saw the advantage of being able to get offshore quickly to conduct rescue operations. Long before there was such a thing as a helicopter that could hoist survivors, a flying boat was the quickest way to get to those in trouble on the open ocean and bring them back alive.

Because Coast Guard pilots were doing this type of flying on a regular basis, they acquired an expertise in open-water aircraft operations that could only come from experience.

For instance, the usual technique for takeoff (or landing) for any aircraft is to head into the wind. This practice increases the flow of air over the wing, which means increased lift while using minimum power and having minimum speed over the surface (i.e., ground speed). The technique works for aircraft on water as well as on land.

However, because wind generates waves, a flying boat that heads into the wind for takeoff or landing is a flying boat that heads directly into oncoming waves. In comparatively light winds or on enclosed bodies of water such as small lakes and bays, this is not a problem as the waves will be minimal.

Yet as wind builds and wave height increases, there *is* a problem. The flying boat will begin to pitch with the waves causing variations in its speed or, at worst, it will catch a wave and nose over, stopping the flight, damaging the aircraft, and perhaps causing it to sink.

There exists a mitigating phenomenon that enables a flying boat to head into the wind on takeoff or landing in running seas: The wind generally never blows from exactly the same direction for very long periods of time. In fact, as weather systems move across the face of the earth, the winds are almost constantly, albeit slowly, shifting from one direction to another. As waves are created by wind, there are usually multiple wave systems running across the surface of the ocean. Therefore, the previous wind will have created a swell (primary sea) that is often coming from a direction that is at an angle to, even perpendicular to, the existing wind and smaller waves (secondary sea).

Consequently, if a flying boat can head directly into the existing wind (and its comparatively smaller waves), it may be able to put the primary sea off to one side of its heading. Thus, if the timing is right, a flying boat can take off or land between and parallel to a pair of large swells. This allows a takeoff or landing that generates maximum lift, minimum ground speed, and little impediment from the surface of the ocean. It means the flying boat can land at the slowest speed and with the least possibility of nosing over and damaging itself. It may not be easy and it may not be smooth, but it will work.

Learning all this, the Coast Guard aviators taught the cuttermen in the ocean station fleet that any pilot not familiar with water landings and with little time to make a decision about a ditch heading should be told to land his aircraft "across the primary and into the secondary."[1] This would result in a landing into the wind and parallel to the major swell system.

The advice from lessons learned over the decades by Coast Guard aviators was incorporated into a manual titled *Aircraft Emergency Procedures Over Water,*[2] which included ditch procedures and was an essential volume in the library of every ocean station cutter.

IN JANUARY 1953, a PBM from the U.S. Coast Guard Air Detachment at Sangley Point in the Philippines took off to rescue the crew of a Navy P2V land-based airplane that had ditched at sea.

The Navy airplane ditched because it had been shot down off the coast of China during the Korean War. The Coast Guard PBM had a pilot, Lt. John Vukic, a co-pilot, Lt. (jg) Gerald Stuart, and a crew of six: Chief Petty Officers Winfield Hammond and Joseph Miller, Jr., plus Petty Officers Joseph Bridge, Robert Hewitt, Tracy Miller, and Carl Tornell.

While flying toward the position of the ditching, the Coast Guard aircrew got good news and bad news. The good news was that other airplanes were on the scene, could see survivors in the water, and had dropped rafts and survival equipment to them. The bad news was that the survivors had gotten none of the gear because the wind was blowing thirty knots and the seas were running eight to twelve feet.

Once they got on scene, Lieutenants Vukic and Stuart looked over the situation and found that seven survivors had struggled into a life raft while four others hung onto the side of it. No one knew when or if a surface vessel was coming to pick them up. Plus, it was getting dark. So the aviators decided to land their flying boat, pick up the survivors, and attempt to take off again.

The PBM touched down with no apparent damage. The Coast Guard crew managed to get the eleven Navy men into the flying boat and then learned that there were still two more missing. Lieutenant Vukic taxied around searching for them. The PBM was pounding badly, and some of the recovered men had injuries that called for prompt, extensive medical care. He would have to take off.

The PBM was equipped with a device known as JATO (jet-assisted take off), which consisted of very-short-duration jet engines. Some propeller-driven military aircraft in the early 1950s used JATO to reach takeoff speed on land more quickly, and it was a reasonable technique to help a flying boat get off the water and into the air as quickly as possible.

As the Coast Guard PBM started its takeoff, however, an engine failed, and the aircraft pitched hard to the left and crashed into the big waves. Now the eleven survivors had become nineteen. Even though injured, Lieutenant Vukic was able to climb into a partially inflated raft and recover two of the Navy men who had just survived their second open-sea aircraft crash in one day. Meanwhile, Chief Miller and Petty Officer Hewitt retrieved a second raft and rounded up five more of the Navy men.

The airplanes that were still circling overhead began dropping more rafts, but the wind carried them away or the seas capsized them or the men, exhausted and injured, simply could not get in them. Those who had gotten into the two rafts rode the cold, sickening ride for two hours until the USS *Halsey Powell* picked up the raft with Chief Miller, Petty Officer Hewitt, and the five Navy men. Three hours later, the crew of USS *Halsey Powell* found the raft with Lieutenant Vukic and the two Navy men.

All night and into the next day the searchers kept looking. They found no one.

Six Navy men had died and five Coast Guardsmen had died trying to save them. It was a noble effort by the Coast Guard aircrew, and each member was awarded a Gold Lifesaving Medal. The effort also demonstrated some of the hazard in conducting open-water rescue with flying boats.

FLYING BOATS ARE USUALLY high-wing aircraft; their wings run across the top of the fuselage. Otherwise, the aircraft could never take off again because the wings would be in the water. To keep the flying boat balanced once it is on the water, it has floats, or sponsons, mounted under the outboard ends of the wings. Otherwise, it would roll over. These factors are all taken into consideration when designing a flying boat.

Designers of land-based aircraft do not have to consider such things, and so they may place the wings either on top of (high-wing), in the middle of (mid-wing), or below (low-wing) the fuselage. When a land-based aircraft does land on the water, however, the position of the wings becomes crucial.

Assuming it lands straight and level, a low-wing airplane on the water has not just the fuselage but also the wings themselves touching the surface of the water. This spreads the landing impact over a greater area of the aircraft.

Any ditching aircraft is going to be slowed down rapidly by the substantial drag of the airframe on the water's surface. The high forces generated by this braking may cause the fuselage to break up, yet the dispersion of these forces over a greater surface area in a low-wing airplane may help keep the plane in one piece.

Finally, after landing, a low-wing airplane will sit (at least for a while) in a relatively stable position because the wings themselves prevent the plane from rolling over. This will give those aboard extra time to get out of the airplane and into rafts or the water before the plane, inevitably, sinks.

A high-wing or mid-wing land-based airplane has none of those advantages. Because the full impact of landing has to be taken by the comparatively small surface area of the fuselage, there is an increased probability of structural damage. And because the wings are not touching the surface as the airplane lands, there is less stability, and the plane is much more likely to either roll over or nose over. Even if the airplane does land successfully on its fuselage, it is inevitably going to roll to one side or the other as soon as air flow over the wings stops. As soon as it rolls, water may start to enter through doors or windows. Thus, of the three types, a low-wing airplane is most likely to survive a ditching long enough for everyone to get out.

The successful rescue of the people aboard *Bermuda Sky Queen* in 1947 was largely due to the aircraft being a flying boat, designed to deal with the marine environment. It not only landed without damage but also stayed afloat for an entire day.

As the 1940s became the 1950s, however, fewer flying boats were being used by either commercial airlines or the military, and more over-water flights were being made with conventional land-based airplanes. When and if other ditchings were to occur on ocean stations, it seemed likely that the cutters would have to deal with aircraft that had little inherent buoyancy and were much more likely to sink shortly after impact. The Coast Guardsmen would not have hours and hours to finish the rescue; they would only have minutes.

And that was the case.

ON APRIL 27, 1949, a U.S. Air Force C-47 (Military Air Transport Service [MATS] 43-16396) was forced to ditch in the dark on Ocean Station Dog (later Delta), which was being manned by the crew of the cutter *Sebago*. The C-47 was a low-wing, land-based airplane, and the conditions were poor with northeasterly winds of twenty knots and seas running ten to twelve feet. The crew of MATS 396 encountered engine problems at 0130, finally deciding at 0152 to ditch. The C-47 touched down at 0251 close aboard *Sebago*. The cutter's boat was away within a minute and alongside the airplane by 0254. The boat crew quickly recovered all four men before the airplane sank at 0305. Seven minutes later, the boat was hoisted, and the survivors were aboard *Sebago*. The total elapsed time from ditching until all the survivors were safe on the cutter was twenty-one minutes.

Sebago's captain, Cdr. Eugene A. Coffin, Jr., took the Air Force men to Argentia, Newfoundland, arriving the next day. At 1710 on April 28, *Sebago* headed back for Ocean Station Dog.

THE CUTTER *COOS BAY* had only been on Ocean Station Echo for ninety minutes when potential catastrophe flew her way. *Coos Bay* had relieved *Duane* at 1041 on February 27, 1953, in the typically sloppy conditions of a mid-winter North Atlantic; southeasterly winds of thirty to thirty-five knots and seas running twelve to fifteen feet.

Just after noon that day, Lt. Cdr. G.M. Battle, the pilot of a Navy P2V patrol airplane (a mid-wing aircraft with the call sign N9357), reported he had lost one of his two engines and was losing altitude some fifty miles east of Echo. Within the hour, Commander Battle informed the crew of *Coos Bay* that he was headed for the cutter and intended to ditch. Twenty minutes later, the cutter crew had the aircraft in sight, and the crews of both ship and plane made final preparations for the ditching. The Navy airplane circled until 1329 when the pilot informed the cutter crew that he was going to radio silence so that he and his co-pilot could concentrate on the final approach and landing.

Coos Bay's captain, Cdr. John P. Latimer, had rigged out two of his cutter's small boats, one each to port and starboard, with the crews in them ready for launch as soon as the airplane landed.

By then the winds were up to forty knots from the southeast, and the swells were shifting from the southwest to the southeast. This left the pilot little choice but to pick a compromise heading that put his airplane's nose into the large, primary swell. The Navy plane touched down at 1331, and two minutes later the violent pitching forces broke the fuselage in half just behind the wing.

However, *Coos Bay*'s starboard boat was already on its way having been lowered within one minute after the aircraft struck the heavy seas. The boat crew could see that six men had gotten into a raft, but the other four who had been in the airplane were now in the water. The boat crew made an instantaneous decision: They first picked up the four immersed men knowing that these men were in the greatest peril.

Twelve minutes after the airplane had touched down, *Coos Bay*'s boat crew had returned those four men to the cutter and departed immediately for the raft. Another fifteen minutes and the six men had been removed from the raft and brought aboard the boat, taken back to the cutter, and were safe on her decks. The total elapsed time from ditching until all the survivors were safe on the cutter was twenty-seven minutes.

Coos Bay's boat crew brings a Navy P2V crewman alongside.

The boat crew searched for debris for a short time before being hoisted aboard. Then *Coos Bay* headed for the cutter *Barataria*, which was sent from Bermuda to pick up the ten survivors. The ten Navy men were all successfully transferred to *Barataria* on March 1, and *Coos Bay* returned to Ocean Station Echo.

TWO YEARS LATER, *Coos Bay* was back on Echo in January 26, 1955, when an Air Force C-54 (MATS 45569) lost two of four engines on a flight from the Azores to Bermuda. The Air Force crew first called the cutter at 1835 reporting an uncontrollable fuel loss from one engine. Within an hour a second engine failed, and the pilot, Capt. Paul Evans, informed the crew of *Coos Bay* that he was going to have to ditch. Conditions were every bit as bad as they had been for *Bibb*'s rescue of the *Bermuda Sky Queen* in 1947 with forty knots of wind and seas running fifteen feet.

But *Bermuda Sky Queen* had been a flying boat; the C-54 was not. *Bermuda Sky Queen* had enough fuel to stay aloft until daylight; the fuel lost by MATS 569 took that option away. It would have to be a ditch in the dark.

The crew of *Coos Bay* made preparations for the airplane's touchdown and, once the pilot said he was on his final approach, began firing illumination flares from a mortar to light up the heaving sea. At 2005, MATS 569 was on the water. Two minutes later, five of the crewmen were in a raft, and the other three crewmen were standing on the exterior of the fuselage. Two minutes after that, *Coos Bay* was close aboard the airplane, and the cutter crew launched a small boat. By 2016, the three crewmen on the fuselage were in the boat, and three minutes after that the five crewmen from the raft joined them. By 2042, the boat was back on deck with everyone saved. The total elapsed time from ditching until all the survivors were safe on the cutter was thirty-seven minutes.

Commander William S. Vaughn, *Coos Bay*'s captain, made the wise decision to clear the area as soon as everyone was safe. In the darkness, howling wind, and driving spray, no one could confirm that the aircraft had sunk, and Captain Vaughn did not want to find the fuselage with his cutter's hull.

The next morning there was no sign of the aircraft, and *Coos Bay* headed toward the edge of Echo where she was met by the cutter *Duane*, and the unexpected passengers were transferred to *Duane* on January 29. *Coos Bay* then returned to Ocean Station Echo.

THERE ARE NO RUNWAYS in the middle of the ocean, no "ground" traffic control, no taxiways, and no nearby obstacles. There are no painted markings, no lights, and no instrument landing system (ILS). There is an almost complete lack of anything to give a sense of perspective about altitude or distance. The surface of the sea looks the same to a pilot at 2,000 feet as it does to a pilot at 1,000 feet. And on a moonless or overcast night, the surface of the sea looks the same no matter what the altitude: It is a uniform black mass. Day or night and especially for those who lack experience flying at low altitudes over water, it is extremely difficult to "sense" how near or far away the surface is.

Consequently, the crew of an ocean station cutter had to provide or improvise all these things (runway, lighting, distance information) for any airplane that was going to ditch. To minimize improvisation and to maximize success, the Coast Guard developed a standard set of procedures to be followed in the case of a ditching. These procedures were all designed for the worst case: a ditching at night. The reasoning was (and is) that if one thought about everything that could go wrong under the worst circumstances, then nothing should have been overlooked for more benign conditions.

Even before an airplane ditched, a cutter crew could provide the aircrew with the latest surface weather including wind and sea conditions and give the pilot a recommended ditch heading. Because the cutter's crewmen were constantly observing and reporting the weather, they could continuously update the ditch heading, and they kept it posted in the combat information center (CIC) so that it was always available on a moment's notice.

The cutter crew could create an artificial runway by laying out a "landing lane" by either spreading firefighting foam on the water (in daylight) or by putting out a string of waterlights (at night). These were lights, not flares, to minimize the possibility of fire in case the aircraft's fuel tanks ruptured during the ditch. The landing lane was laid out along the ditch heading, which was based on the existing weather conditions. Because the ditch heading was constantly updated in the CIC, the cutter crew could, in extreme circumstances, begin laying out a ditching lane within minutes.

During a night ditch, a cutter's guns or mortars could provide illumination of the water surface to give the pilot the visual reference that was otherwise lacking. First, the cutter crew would lay out the landing lane. Then, the cutter would be stationed near (but not at) the end of the lane and off to the left side. As the aircraft began its final approach, the cutter crew would fire parachute flares beyond the end of the lane.

This would illuminate the surface without blinding the pilot and would also ensure that the flares would land in the water well beyond the aircraft, once again to minimize the risk of fire.

The cutter's air-search radar allowed the crew to accurately determine the airplane's position. Thus, the cutter's CIC team could act as *de facto* air-traffic controllers and provide courses to fly, distance to touchdown, and recommended altitudes to an aircrew that was going to ditch its aircraft.

Finally, the cutter crew would launch one or more of the small boats to get to the airplane as quickly as possible after the ditch was completed. Maintenance of speed and efficiency was one reason the cutter crews needed to practice boat launching.

Practice was also required to maintain proficiency in the entire gamut of skills required for successful ditching. The Coast Guard mandated annual training requirements for ditch drills using Coast Guard aircraft, and this was one purpose served by the Coast Guard aviation detachment at Bermuda.

The Bermuda SAR Standby served multiple purposes: It kept a cutter on search-and-rescue standby; it effectively put a cutter and the Coast Guard aircraft several hundred miles "at sea," which could reduce transit and flight times for SAR missions; and it enabled cutter crews to get direct training in ditch and rescue.

Not all of the ocean station cutters were on the East Coast, however, and thus the cutters on the West Coast had to get their ditch-and-rescue training in the Pacific. Consequently, the crewmen of the *Owasco*-class cutter *Pontchartrain* had gotten their most recent training in San Diego in August 1956. *Pontchartrain*'s captain, Commander William K. Earle, would later describe this training as "valuable" for giving the entire crew "an immediate appreciation of the many factors involved" in a real-life ditching.

ALL THE ELEMENTS—the existence of the ocean stations, the experience in aircraft operations on open water, the development of and actual experience in using ditch procedures, and the continual training—came together on the evening of October 15, 1956, when the

Pan Am clipper *Sovereign of the Skies* (tail number N90943), a four-engine, low-wing Boeing Stratocruiser, left Honolulu on its continuing flight from Tokyo to San Francisco. Additional passengers had boarded the aircraft in Honolulu, the plane's fuel was topped off, and there was no reason to suspect problems because the aircraft had performed well thus far on its trans-Pacific journey.

The plane reached its cruising altitude of 21,000 feet and was proceeding normally when the no. 1 propeller sped out of control. The crew struggled with the vibrating aircraft and finally got the engine stopped, but the pilot, Captain Richard Ogg, had lost 25 percent of his power in midocean. He immediately radioed Ocean Station November and informed *Pontchartrain* that he might have to ditch. It was 0322 local time on October 16.

Once a pilot elects to ditch, he is admitting that he is at his final option. So Captain Ogg did not commit himself to ditching when he first radioed the cutter. Instead, he asked for a continuous broadcast of the cutter's radiobeacon so that he could home on it and thus pass directly over the cutter and that the cutter crew be prepared for the possibility of a ditch.

Three minutes after the radio call, when *Sovereign of the Skies* was about forty miles away from *Pontchartrain*, the airplane's no. 4 engine failed. Now with half his power gone, land of any sort more than a thousand miles away, and darkness all around, Captain Ogg asked *Pontchartrain* to make all preparations for ditching.

The cutter's crew had already started those preparations when the aircraft made its first call. One of the cutter's small boats was rigged out ("at the rail") and manned but not lowered, deck rescue gear (ladders, lines, life rings) was made ready, and the electric waterlights were broken out and taken to the fantail in preparation for deployment.

The Stratocruiser passed overhead at 0335, and the pilot notified the cutter that he was going to fly a short distance away and dump fuel as a precautionary measure before ditching. As he did so, he found that he could still hold altitude at about 2,000 feet. The aircrew continued to attempt to get the stopped engine started again. If this could be done, then the aircrew might attempt to make the West Coast or, alternately, return to Honolulu. But the aircrew also knew that without three operating engines, any attempt to proceed or turn back would be foolish because the aircraft might then be forced to put down in the ocean with no ships at all nearby.

THE AIRCREW HAD no luck with the recalcitrant engine, and so much fuel had been burned by 0500 that flying anywhere else was no longer possible. Thus, rather than compound his dire situation by ditching in darkness, Captain Ogg agreed with *Pontchartrain*'s Captain Earle that the actual ditch should wait until daylight.

Although he was an experienced Coast Guard officer, Captain Earle was making his first patrol of an ocean station as the cutter's commanding officer. Throughout the remaining hours of darkness, he and the pilot discussed all the possibilities and necessities involved in ditching the airplane and saving its passengers and crew. The pilot provided precise information about the airplane's exits as well as advice on how best to open them should they become jammed during the ditch. He relayed that the airplane had two life rafts onboard and that he would have his crew deploy them at once after the airplane stopped moving. Captain Earle explained how his cutter would put down the landing lane and that he would launch small boats to get the passengers and crew away from the airplane promptly and then onto the cutter. They had discussed, they believed, every possible aspect, and they waited the additional minutes for daybreak.

Meanwhile, the airplane's passengers had been given complete instructions about what was happening and what they would have to do and not do when the airplane ditched. All the passengers were moved to the forward part of the cabin near the wing area. This should have been, and proved to be, the point of greatest strength for the fuselage and would thus give the passengers the best chance of survival. As dawn broke, they were given final instructions. There was no panic in the airplane's cabin, and the passengers were as confident as the pilot that all would turn out well.

The airplane stayed aloft, waiting for more light, burning off the last of its fuel to minimize weight and to decrease the possibility of fire. The weather was favorable: The wind was a zephyr, the seas were

Pontchartrain's crew lays down a foam landing lane.

calm, and it was a clear day. Finally, *Sovereign of the Skies* made its approach to its ultimate landing as the crew of *Pontchartrain* laid down the foam runway.

The aircraft struck the water smoothly and bounced once. Shortly thereafter, it nosed over and the tail broke away proving the wisdom of moving all the passengers to the over-wing area. *Pontchartrain* steamed up to the stricken craft at full speed ahead and then backed full as the crew lowered two boats into the water. The passengers and crew appeared on the plane's wings within two minutes after the ditching and launched the airplane's rafts. Within seven minutes after touchdown, the first survivors were being brought aboard one of *Pontchartrain*'s boats. The cutter's second boat dashed over to the airplane, and one of *Pontchartrain*'s crew, Petty Officer Christian Ronald, volunteered to enter the sinking, broken airplane to make sure no one was left inside. Once he finished his search, the second boat began taking the remaining passengers and crew aboard.

Thirty minutes after touchdown, all thirty-one people were aboard *Pontchartrain*. The boats stayed in the water picking up luggage, mail, and whatever else floated free. It had been a textbook example of a ditching under ideal conditions.

Even though there was no cutter on scene for relief of station, *Pontchartrain* was immediately released from station to take the survivors to San Francisco. The Coast Guard would not subject a group of civilians to a lengthy stretch on station, particularly when no facilities were available to adequately accommodate them. *Pontchartrain*'s offic-

Sovereign of the Skies ditches.

Pontchartrain's boat crew to the rescue as the fuselage of *Sovereign of the Skies* lies broken on the sea.

Pontchartrain arrives in San Francisco with the survivors from *Sovereign of the Skies.*

ers and chief petty officers gave their quarters to the survivors and either slept on cots or mattresses in other spaces or moved into empty racks as the cutter made the week-long transit to San Francisco.

BY 1956, DITCHINGS had become infrequent, and it was the rarity of the event that made the ditch and rescue of *Sovereign of the Skies* newsworthy—no large commercial airplane had ditched on an ocean station in nine years. *Life* magazine even featured the story on its cover. Photographs of the ditch and subsequent rescue became instant "classics" and were later used in Coast Guard publications to demonstrate how a ditch and rescue should be conducted. It became, literally, a textbook example of ditch and rescue.

In the decade after World War II, five ditchings took place on ocean stations, and a total of 122 people were saved with no loss of life or serious injuries. The conditions had ranged from ideal for *Sovereign of the Skies* to terrible for MATS 45569.

But the Coast Guard's record was consistent and, just like the performance of *Pontchartrain*'s crew, perfect.

IN THE PERIOD between *Bibb*'s rescue of the *Bermuda Sky Queen*'s passengers and crew (1947) and *Pontchartrain*'s somewhat simpler but no less successful rescue of those aboard *Sovereign of the Skies* (1956), the United States went to war in Korea. Once again, the nation found it had to send personnel across the expanse of the Pacific Ocean.

And once again, the Coast Guard was called on to expand its mission and provide the necessary weather, navigation, communication, and search-and-rescue services to aircraft flying from the West Coast to Japan and Asia.

Thus, additional Pacific ocean stations were established beginning in September 1950 with Station S (Sugar; 48°00' N, 162°00' E) followed in 1951 by Stations Q (Queen; 43°00' N, 167°00' W), U (Uncle; 26°00' N, 149°00' W), and V (Victor; 33°00' N, 164°00' E). Of the two Pacific ocean stations extant prior to the expansion, Station N (Nan, later November; 30°00' N, 140°00' W) remained active whereas Station O (Oboe; 40°00' N, 142°00' W) was closed.[3]

The sudden need for additional cutters to patrol an increased number of Pacific ocean stations meant that the Coast Guard had to find ships and find them fast.

So the service turned to a logical source: the Navy's moth-balled ships. There the Coast Guard found what seemed to be an answer to the problem: a "fleet" of ships all of the same class (and about the same size as the existing ocean station cutters) that had been manned by Coast Guard crews for escort duty just a few years earlier during World War II. Because Coast Guardsmen already had experience in these ships, training and familiarization time would be minimized. And because the ships were all of the same class, potential problems such as availability of parts would also be minimized.

Thus, twelve U.S. Navy *Edsall*-class destroyer escorts (1,490 tons, 306 feet) were pulled out of their backwater berths, returned to life, fitted out for ocean station duty, and commissioned as U.S. Coast Guard cutters.

These ships were about the same size as the other Coast Guard cutters patrolling ocean stations. Because they were designed as warships, they were very narrow for their length (with a 37-foot beam), which minimized their resistance and thus maximized their speed.

Which was fine for fighting a war but not so fine for battling the elements, because the narrow beam meant they would roll heavily, and their light displacement meant they would pitch (and pound) hard once the weather built up.

During the Korean War, all but one of these cutters would be sent to West Coast and Hawaiian ports to patrol Pacific ocean stations.

THE FITTING OUT of the *Edsall* class included the construction and installation of a balloon hangar on each ship.

Weather balloons were launched from ocean station cutters four times per day. The simple reflective balloon would be launched to determine the winds aloft via tracking of the balloon by a cutter's air-search radar. The balloon's rate of ascent was known, and the horizontal distance that the balloon had traveled in a given period of time was tracked. Thus, wind speed in the spaces between defined altitudes could be calculated. Information about winds aloft was important not only to aircraft flying overhead but also to meteorologists, who used that information to develop their forecasts.

Other factors used to develop forecasts include temperature, humidity, and barometric pressure. This information was obtained from a device known as a radiosonde, which was a small instrument package including a radio transmitter. For two of the four-daily weather balloon launches, a radiosonde would be attached to the weather balloon. During its voyage aloft, the radiosonde would collect then transmit the data, which would be received by the cutter.

The weather balloons were about four feet in diameter. Inflating such a balloon was no mean feat and would have been impossible on the open deck of a ship of any kind if wind was blowing. Simulta-

Edsall class: The cutter *Richey*.

neously, it would have been equally impossible to inflate such a balloon inside a typical compartment of a ship where the widest opening to get on deck is usually no more than two feet wide. A four-foot balloon simply will not fit through such an opening. Therefore, the ocean station cutters were fitted out with balloon hangars (or balloon shacks) for the purpose of inflating and launching these balloons.

The hangar was located aft of amidships (to prevent a balloon from fouling itself on masts or antennae), above the main deck (usually on the 01-deck level), and was a tall structure with a "roll-up" door in the aftermost bulkhead. When inflation was in progress, the door would be shut. When launch time came, the door would be rolled open, and one or two crewmen would take the balloon (with or without radiosonde) out the door, over to the leeward side of the deck, and release the balloon.

THE PACIFIC OCEAN STATIONS were far-flung, which resulted in long transit times no matter where a cutter was berthed. Because of the stations added during the Korean War, the Pacific ocean station cutters during the early 1950s often conducted multiple patrols with only abbreviated in-port periods in Japan to take on fuel, food, supplies, and some well-needed rest. The requirement to stand search-and-rescue standby duty in Adak (the Aleutians), Guam, or Midway in addition to usual duties gave the Pacific ocean station crews of the period long deployments.

On September 28, 1952, the *Owasco*-class cutter *Winona* sailed from Port Angeles bound for Ocean Station Victor, which was located between Hawaii and Japan. The cutter was scheduled to return home just before December 25 almost ninety days later.

Winona's crew spent three weeks on Victor then sailed to Yokosuka, Japan, where the cutter took on the usual full load of food and fuel. *Winona* then sailed for Ocean Station Sugar, located about 300 miles southeast of the Kamchatka Peninsula. After three weeks on Sugar, *Winona* was relieved on December 13 and headed home, arriving at Port Angeles at 0952 on December 23.

But one hour after arrival, *Winona* was ordered to get under way. Offshore, in thirty-knot winds and building seas, the freighter *Maple Cove* was disabled with propulsion failure. *Winona*'s crew loaded fresh provisions and sailed at 1406.

Rolling through the seas, *Winona* arrived at *Maple Cove*'s position twenty-five hours later and stayed with the stricken ship all night and into the following day. At 1710 on December 25, the commercial tug *Island Sovereign* took *Maple Cove* in tow. *Winona* continued escorting the tug and freighter until 0415 on the following morning, when she was released from the case. The cutter arrived home, again, at 0900 on December 26. Christmas would come late that year.

AFTER THE JULY 1953 armistice brought an uneasy peace to the Korea Peninsula, Stations Q, S, and U were closed, and the *Edsall*-class cutters were decommissioned by June 1954. Stations N and V would continue to be patrolled into the 1970s, yet another Pacific ocean station would exist for a brief period.

The press release announcing the establishment of Ocean Station Romeo (49°00' N, 169°00' W) on August 31, 1958, described it as being "approximately 1,260 miles NW of Honolulu."[4]

Romeo was indeed more than 1,200 miles away from Honolulu. But that also meant it was only about 300 miles south of the Aleutian Islands—not known for balmy breezes and swaying palm trees. Romeo was close to the great circle route from Seattle to Japan, and the station's latitude was less than 250 miles south of Charlie and 300 miles north of Delta. In those latitudes, the Pacific Ocean is better known for cold, fog, and sudden, extreme changes in weather. The winds shift direction rapidly and build quickly when fast-moving low-pressure centers sweep from west to east just as they always do in the higher latitudes.

On December 17, 1958, the *Owasco*-class cutter *Klamath* was to be relieved as Ocean Station Romeo by the *Casco*-class cutter *Gresham*. *Gresham* radioed that her crew had mail to send back, and *Klamath* agreed to take it. The two crews intended to transfer the mail by running a line from *Klamath*'s forecastle to *Gresham*'s stern, depositing the mail inside a waterproof canister, and hauling the canister from one cutter to another.

This evolution was complicated by the weather. When *Gresham* arrived on station at 0130 on December 17, the wind was blowing out of the west-southwest at forty knots, so the crews decided to wait until daylight to do the transfer. At about 0800, the wind dropped to twenty knots, and the crews decided to make the transfer. *Klamath*'s crew fired a line over to *Gresham*. The sea, however, was still sloppy, and the motion of the two cutters caused the line to part. *Klamath*'s crew fired again, but that line parted as well. So did the next line, as did two more lines.

This was not working, so *Gresham*'s crew started to rig up and lash together two canisters supported by life jackets and wood, which could be floated over to *Klamath*. By now it was after 1000, and the wind that had "diminished" for two hours built again to almost fifty knots and would stay there for the rest of the day.

Gresham's crewmen were working on the large, flat, wide-open aft deck of their cutter. These decks had been perfect when the ships had been seaplane tenders during World War II, and they continued to be useful as an uncluttered work space when handling lines or towing. But the very things that made these such good work areas (lack of bulwarks, low freeboard) left the crews on those decks exposed to any seas that surged aboard.

Just after 1100, one wave (later described by an officer as "tremendous") inundated *Gresham*'s stern. The water crashed down on twenty-two men, throwing them across the deck and into bulkheads, bitts, stanchions, and each other. Legs, arms, ankles, and knees were broken. Backs, faces, hands, and legs were dragged across the deck where they were lacerated, abraded, and bruised. Broken, bleeding, and soaked by 39°F water, twenty-one of the men struggled back to their feet or were helped off the deck by shipmates, who then carried them to sick bay.

As all the tumbling and crashing and shattering of limbs was going on, one young officer, Ens. George T. Bergman, was knocked down and struck his head on something, and the cascading water carried him over the side. His shipmates tossed life rings and life jackets toward him as *Gresham* struggled to get alongside and recover him.

In the chaos on the back deck, no one could be positive that only one man had gone overboard. *Gresham*'s captain, Cdr. Benjamin D. Shoemaker, ordered an immediate muster of the crew, which confirmed that Ensign Bergman was the only man missing.

Simultaneously, *Klamath*'s crew threw life rings and life jackets toward the unmoving man who was sometimes seen and sometimes lost in the tumultuous seas.

Gresham's log entry at 1359 for that day reads, "Ensign George Thomas Bergman, USCG presumed to be lost at sea. May God have mercy on his soul."

Fifty days after Ensign Bergman's death, Romeo was closed and never reopened.

ENDNOTES

1. Conditions at sea are never as simple as this axiom suggests. Standard ditch procedures for military pilots included a number of caveats and alternatives such as accepting a substantial crosswind in order to parallel major swells, or landing parallel to the primary sea and *with* the secondary sea if the secondary was large, or possibly ignoring the swell direction altogether and landing into the wind if the wind was very strong even if that meant heading into the primary swell.

2. Smith (1968). The principal author of *Aircraft Emergency Procedures Over Water* (CG-306) was Capt. John M. Waters, Jr., USCG.

3. See discussion in Chapter 2.

4. U.S. Coast Guard Commander, Fourteenth District (1958).

9

VICTOR

34°00' N, 164°00' E

MIDWAY IS THE quintessential Pacific atoll. Low-lying and studded with palm and pine trees, it sits as an opalescent oasis in the sapphire-blue expanse of the North Pacific. The tranquil shores of its small islands and its reef beckon those making transits of the world's largest ocean. It can serve as an idyllic respite from a long ocean journey.

It can also be a ship-killing trap.

Midway is at the northwestern extremity of the chain of dormant volcanoes that make up the Hawaiian Islands. Once these volcanoes stop erupting, they begin to succumb to the wind and waves of the Pacific Ocean until they are finally eroded to a collection of rock and sand mounds, some just above the waterline and some just below it. At the same time, coral begins to grow on the submerged mountainsides so that a wide yet short, hard circle just breaks the ocean's surface. These coral, limestone, and volcanic rock formations are lethal to the hulls of ships.

Two nineteenth century British whaling ships, the *Pearl* and the *Hermes*, gave their names to one such atoll about 100 miles east-southeast of Midway. The ships ran aground on the atoll in 1822 and were destroyed, but their crews cobbled the pieces together into a "new" small ship and sailed it to Honolulu.

French Frigate Shoals is another such atoll located about 550 miles southeast by east of Pearl and Hermes Atoll. The crews of the French frigates *la Boussole* and *l'Astrolabe* discovered the atoll in a nearly miraculous sighting of the shoal's breakers at 0130 on November 6, 1786. Though the French ships avoided striking the reef (by a scant 200 yards), the U.S. vessels *South Seaman* and *Daniel Wood* would both fetch up on the shoals and be destroyed in the middle of the nineteenth century.

Despite their isolation and risk to ships, these atolls are useful. It is possible to construct airplane runways and buildings on them. Tern Island, for instance, a part of French Frigate Shoals, had a runway built on it as a way station for airplanes bound for Midway from Hawaii during World War II. Later during the war, a loran station was built adjacent to the runway, a station that would serve navigators for more than three decades. Furthermore, it is also possible to use the interior portion of an atoll—the remnant of the volcano's caldera—as a safe anchorage area. If an entry exists or is blasted through the surrounding rock and coral, the middle of the atoll may provide a deep, well-protected anchorage.

Midway thus had strategic importance. Located some 1,140 miles northwest of Honolulu, Midway is effectively a way station on the route to Asia. Consequently, a Pan American Airways base, a Naval Air Station, a submarine base, and a Naval Operating Base were built at Midway.

THE *OWASCO*-CLASS cutter *Iroquois* approached Midway on the morning of June 29, 1954, to take on fuel while headed for Ocean Station Victor. Even though *Iroquois* had left Honolulu with 117,000 gallons of fuel, her crew intended to take on more at Midway for a very good reason: The cutter had already steamed almost 1,140 miles and still had another 800 miles to go before she reached her destination.

Ocean Station Victor was a long way out in the North Pacific. Victor was so far out it was not even in the same hemisphere as the other ocean stations: It was in east longitude. Victor was so far out that Tokyo was 1,500 miles closer than San Francisco. Victor was so far out that Yokohama was 500 miles closer than Honolulu.

This geographic fact helped drive the operational scheduling for Victor. Because it took less time to get to Yokohama than it did to reach Honolulu, cutters, especially West Coast cutters, would spend a typical three weeks on station, then sail to Japan for fuel and other logistics, return to Victor for another three weeks, and then head home. This became known as a *Double Victor.*

Double Victors were both good news and bad news: The good news was that ocean station cuttermen, normally accustomed to seeing nothing but their homeports and a whole lot of open ocean, got to make visits to Japan. The bad news was that they were gone from home for as much as three months at a time.

The cutters homeported in Honolulu were more likely to make "single" Victors and in fact would sometimes serve as the midperiod relief cutter for a West Coast cutter on a Double Victor.

It was in any cutter's best interest to arrive on Victor with a healthy load of fuel and provisions because both "gas and groceries" were a long way away. Thus *Iroquois* began slowing down at 0525 on June 29 as her crew made preparations to enter Midway's channel and moor at the Navy base to take on fuel.

Iroquois' crew made all the usual preparations any ship makes for entering restricted waters: Both of her steering gear engines were engaged so that if one failed, the rudder could still be controlled; extra lookouts were posted and additional navigation personnel were brought to the bridge; and her captain, Cdr. Ashton H. Barnes, arrived on the bridge to oversee the operation of his cutter. *Iroquois* maneuvered off the channel entrance to pick up a ship's pilot from a Navy tug. At 0748, the pilot, a U.S. Navy warrant officer, came aboard and went directly to the bridge.

THE ROLE OF A ship's pilot is often misunderstood. A pilot is an advisor: a heavily respected advisor who has extensive knowledge about the waters the ship is transiting. Pilots must have a great deal of experience in the waters on which they serve and, at least in the United States, are extensively tested to ensure their expertise. Unlike the ship's crew, which may see a port only once every few months, most pilots are transiting the ports they serve on a daily basis. Pilots know the waters:

They know the tides, currents, depths, channels, shallows; they know the most recent changes in the waters; and they know what will and will not work. The pilot is someone who is expected to give the ship's captain and the ship's officers prompt and accurate information about where to take the ship to complete the transit successfully.

Nevertheless, the pilot is still only an advisor. With very few exceptions (such as the Panama Canal), the pilot is not "in charge" of navigating the ship. The pilot does not—cannot—relieve the captain of the ultimate responsibility of command. Still, the pilot is often given the ship's conn; that is to say, the pilot gives orders to the helmsman and engine room to control the ship's direction and speed. It is simply easier, most of the time, to have the pilot give engine and steering orders directly rather than waste time and risk confusion by having him or her make recommendations to the captain or watch officer who then executes them.

So it was aboard *Iroquois* that morning. The pilot ordered standard speed, about thirteen knots, as the cutter passed the first buoy headed in. The cutter was in the channel and on a safe heading. The pilot ordered left 5 degrees of rudder, and the helmsman complied. The cutter started to swing to port, as ordered. That course change, however, would start to take *Iroquois* out of the channel to port rather than heading the cutter straight in. Five seconds later, the pilot ordered left 10 degrees rudder, and the helmsman complied. Now *Iroquois* started to swing more rapidly to port, heading her further out of the channel. The pilot immediately ordered left 15 degrees rudder and then ordered the engine back full.

The backing bell was a desperate measure to avoid grounding the cutter, but it was too late. Backing a ship's engine is not as effective as applying the brakes in a car. Propellers are designed to push ships forward; they are horribly inefficient at stopping them or backing them up. Compounding the problem is simple physics. *Iroquois* displaced 1,980 tons. She was going ahead at thirteen knots. There was a lot of inertia in that mass, and it would take a relatively long-term application of substantial power to stop it. Once the rudder orders took the ship out of the channel, there was less than fifteen seconds in which to stop her before she fetched up on the rocks and coral of Midway. There was simply not enough time for the backing bell to prevent catastrophe.

The series of left rudder orders had made no sense, but they had come so quickly that Captain Barnes had no time to counteract them. Less than a minute after the pilot's first ill-advised left rudder order, *Iroquois* was hard aground. Her engines were still backing full but she was stuck.

At 0805, the Navy tug that had brought the pilot out put a line on *Iroquois*' stern and attempted to help pull her off while the cutter continued backing full. The tug, which had no name only the number YTB379, pulled so hard that after seven minutes, the wire towing cable snapped. The end of the cable splashed into the water and immediately fouled itself around *Iroquois*' propeller. The engines were stopped, and the crew managed to get the line out of the screw.

A ship is no small object, and whereas one part of it may be aground, the rest of it, even most of it, may still be afloat. So the key is to get the part that is grounded back into deep water while not allowing the part that is afloat to become grounded. Consequently, once *Iroquois* got her propeller cleared, she only started backing at one-third speed not so much to refloat herself but to avoid going further aground.

Simultaneously, *Iroquois*' crew took soundings (i.e., measured the depth of the water) all around the cutter to determine which part of her was aground and where she might still be afloat. The news was not good. The soundings showed fifteen feet of water aft, fifteen feet amidships, and ten feet forward. *Iroquois*' draft should have been about seventeen feet aft, fifteen feet amidships, and twelve feet forward. In other words, the cutter was truly hard aground.

Furthermore, the term *hard aground* was accurate. Although Midway has some sand, that sand generally provides only a thin cover over the rocks and coral. If there was any sand where *Iroquois* grounded, it provided very little cushion to the cutter's steel.

AT 0828, THE CREW prepared to lower *Iroquois'* boats. This would decrease the cutter's weight and thus its draft thereby making it easier to float. Furthermore, the boat crews could take a look around the cutter and see if there was any visible damage to the cutter or signs of oil, which might indicate that a fuel tank had ruptured. YTB379 came alongside again, and a 12-inch-circumference manila towing hawser was passed to the tug for another attempt to pull the cutter off.

While the tug with no name was alongside, something inexplicable happened: The Navy pilot stepped off *Iroquois* and onto the tug. He did not have anyone relieve him of *Iroquois'* conn. He did not tell the watch officer he was leaving. He did not tell Captain Barnes he was leaving. He just stepped away from the mess he made leaving it to others to clean up.

This was not the only thing going wrong. The big hawser fouled *Iroquois'* screw. As soon as it was cleared, it fouled the tug's screw. The tug struggled away to clear its own propeller as *Iroquois* tried once again to back herself off the coral with no luck. Now the sea itself turned against the cutter.

The weather was typically glorious Hawaiian: unlimited visibility, cumulus clouds, air and water temperatures both in the upper seventies Fahrenheit, and the usual twenty-knot trade winds blowing.

The trade winds occur in two bands of latitude: a northern band stretching from the Equator to 30° N and a southern band stretching from the Equator to 30° S. The winds of the northern band blow from the northeast toward the Equator and are known as the northeast trade winds, and the winds of the southern band blow from the southeast toward the Equator and are known as the southeast trade winds. Strong enough to propel the heaviest sailing ship without being overpowering, consistent enough in direction to provide day after day of propulsion, the trade winds made seaborne trade in the age of sail not just possible but also reliable. The very strength and constancy of the trade winds means that they build up a continuous sea. Swells are always running in the belts of the trade winds, but because the winds blow over such broad expanses of ocean, the swells stay comparatively far apart and are relatively benign.

Until they approach land. Then, as is true with all waves, the distance between them shortens, their height increases, and they will surge and eventually break against the shore. These swells will bend around an island, and on the side of the island away from the wind, the swells will heave the water up and down. The Hawaiian Islands, including Midway, lie in the northeast trade winds of the Pacific. Thus, each of these trade wind–driven swells would briefly pick *Iroquois* up—not enough to refloat her but just enough to lift her off the bottom momentarily—and then drop her down again. Hard. The trade winds had been blowing for centuries. The swells had been running for centuries. They would not relent. *Iroquois* was picked up and slammed down over and over again.

It was rare for a Coast Guard ocean station cutter to receive a log entry about the ship rolling or pitching or pounding. Only the truly adverse conditions, the ones that caused damage to the cutters or that injured crewmen, were put in the log.

At 1000 on June 29, *Iroquois*' log noted "vessel pounding on coral with heavier swells." At 1246, the cutter's log noted "ship pounding quite heavily." Later, the log would read "ship taking considerable pounding."

The pounding was indeed severe and considerable: It was breaking up the cutter. Eight minutes after the "pounding quite heavily" log entry, a leak sprang in the cutter's gyro room, which was located low in the ship. Internal frames began to buckle. Steel plating all over the bottom of the cutter was being dished in. The keel—the central, critical support member of the ship's framing structure—was bent where it ran through the engine room.

IROQUOIS' CAPTAIN and crew were not, however, going to just stand by and let their cutter get pounded to death. By 1130, the Navy tug was able to retrieve the 12-inch towing hawser again and began pulling as *Iroquois* backed her propeller. Within six minutes, this huge hawser had parted under the strain as had the wire towing cable earlier. It was obvious that *Iroquois* was not going to refloat by use of her own engines and the power of one small tug.

Unable to get off the shoal, with the incessant pounding adding to the damage minute by minute, and with no other help immediately available, Captain Barnes had to reset his priorities. He probably suspected, and would later learn for sure, that the swells were not only pounding the cutter but also pushing it farther up the reef making his already perilous situation worse. He would attempt to stabilize the cutter's situation as much as possible by holding it in place with anchors set offshore by YTB379.

After the towing hawser parted, the tug had returned to the Navy base and picked up a large anchor for this purpose. YTB379 returned to *Iroquois*, attached the cutter's wire cable to the anchor, carried the 5,000-pound anchor 700 feet astern of *Iroquois*, and dropped it. The cutter took a strain on the cable and hoped the anchor would hold.

By 1430, Captain Barnes made another unpleasant but reasonable decision. He began a *de facto*, but not declared, abandonment. The cutter's crew gathered up the personnel records, classified material, and cash. (In the days before direct deposit, cutter crews were always paid in cash, so there was a lot of it aboard each ship.) All of this and some of the crew were put aboard the tug and sent in to Midway's Sand Island. The cutter's small boats began taking the rest of the crew ashore. By 1715, only seven officers and forty-two enlisted men remained aboard. The other seven officers and seventy enlisted men were ashore.

Meanwhile, the Coast Guard's Fourteenth District Headquarters also realized *Iroquois* was not going to get out of her predicament without help and asked the Navy's Pacific salvage command (COMSERVPAC) to assume control of the cutter and situation. They did so at 1550 and sent the Navy's salvage vessel USS *Deliver* to Midway. They also dispatched a salvage officer to the scene by aircraft.

YTB379 set another anchor off the cutter's starboard quarter at 1800. *Iroquois* picked up her boats. Those aboard prepared for a long, pounding night during which they patched holes, reinforced buckled structures, and shored up bent bulkheads.

The following morning, June 30, one of *Iroquois*' small boats went in to Sand Island and picked up the Navy's salvage officer, Cdr. J.P.

Lehan. Once he completed his evaluation in the afternoon, *Iroquois* began getting her crew back aboard. All hands would be needed both to get the cutter afloat and keep it afloat once it was off the bottom.

Deliver arrived at 0624 on July 1. Her crew began deploying anchors and other gear so that *Iroquois* could be pulled off through a combination of kedging and towing.

KEDGING HAS BEEN DONE for centuries to get ships back afloat. Before ships had mechanical propulsion, it was often the only way a grounded ship could be refloated. It is a simple process. A large hawser (or, more recently, a wire cable) is attached to an anchor. A small vessel then carries the anchor offshore and drops it. The ship begins heaving around on the cable by using a strong capstan or a windlass. If the anchor is far enough offshore, the hawser strong enough, and the capstan powerful enough, the ship will literally pull itself back to the anchor and thus into good water.

However, grounded ships sometimes did not have any power available for the capstan or at least no capstan with sufficient "pull" to heave around successfully on the kedge-anchor's cable. Even if they did, the strength of the cables and anchors was not always sufficient for the job. That was the problem with the initial efforts of *Iroquois* and YTB379.

During World War II, the U.S. Navy refined the kedging concept by assembling what became known as beach gear. The first part of the beach gear was an anchor with powerful holding ability that was attached to chain and a heavy wire cable. That cable was then attached to a set of blocks led with wire. The blocks multiplied the pulling force, which was supplied by the final piece of beach gear: portable, self-contained winches. All this equipment was made up in compact units so that it could be carried by a small boat, hoisted aboard the stranded ship, rigged up, the anchor led out, and set to work.

That was part of the plan for *Iroquois*. The rest of the plan consisted of *Deliver* using a tow line to help pull on *Iroquois* as the cutter heaved around on the beach gear. It took all of *Deliver*'s 3,000 horsepower plus *Iroquois* heaving on the beach gear, but finally, at 1645, *Iroquois* was afloat. *Deliver* took her in tow and moored her at Sand Island. *Iroquois'* crew patched up and shored up the cutter to make her ready for towing to Pearl Harbor where she could be dry-docked.

THERE WAS NO good news once the cutter was fully examined. Many, perhaps most of the cutter's bottom hull plates were badly dished in. The propeller blades were nicked. The sonar dome was dented and scraped as was the transmitting head for the depth sounder. The bilge keels, long fore-and-aft pieces of flat steel designed to minimize the cutter's rolling, were bent so badly they looked like ribbons rather than steel. The inside of the cutter was no better: Transverse bulkheads (those running from one side of the cutter to the other) were twisted and broken, athwartship frames were no longer attached to adjacent plating, and longitudinal frames (or stringers, which run fore and aft) were bent and had broken free.

Anything can be fixed if enough money is spent on it. But at some point, the dollars going into a problem exceed the value of the finished project. *Iroquois* was at that point. The Coast Guard simply could not justify what would amount to rebuilding a cutter that was already eight years old. The service decided to lay up the cutter at the Coast Guard Yard in Curtis Bay where she would be available as a source of parts.

That decision meant, however, that the ocean station fleet was now short by one cutter. Another cutter would have to be found to take *Iroquois'* place in the ocean station operations schedule.

At the same time *Iroquois* was being pounded on Midway's reef, the *Owasco*-class cutter *Chautauqua* was arriving at the Panama Canal bound for the Coast Guard Yard from her homeport of Alameda, California. *Chautauqua* had been selected for decommissioning by Coast Guard Headquarters because she was "below an acceptable material condition for continued operations." Her reduction gears were badly worn, several watertight doors were improper, one of the anchor windlasses was frozen (despite a recent overhaul), the gyro compass (also recently overhauled) was troublesome, one of the cutter's electrical generators needed repair, and some of the wiring circuits were defective.

By the time *Iroquois* got off the reef, however, the standard for "acceptable" had been altered: *Iroquois* was now even less acceptable than *Chautauqua*. *Chautauqua* was directed to Hawaii to replace *Iroquois* while the Coast Guard made plans to tow *Iroquois* from Pearl Harbor to Curtis Bay.

Many of *Chautauqua's* regular crew had already been transferred off the cutter; she was being sailed to Maryland by a minimal crew all of whom had orders elsewhere once their delivery passage was complete. Because *Iroquois* was now going to be decommissioned, the logical decision was to transfer *Iroquois'* crew, from Captain Barnes right down to the most junior seaman, to *Chautauqua*. And that's what happened.

When Captain Barnes took over command of *Chautauqua*, he and the officer he relieved, Lt. Cdr. Bernard E. Kolkhorst, prepared a lengthy letter noting all the deficiencies in the cutter. But that list was modest compared with the list of damage to *Iroquois*.

Iroquois arrived at the Coast Guard Yard in September 1954 where she was decommissioned and stored until sold for scrap in January 1965. But her real end had come on that reef in the far Pacific.

And so the only peacetime loss of an ocean station cutter did not occur during a raging midwinter gale in a blinding snowstorm while trying to rescue drowning men. Instead, it happened on a quiet atoll in perfect summertime weather during a routine stop for fuel.

LOGISTICS AND LIBERTY in Japan were not the only things that drew Coast Guard ocean station cutters to the Far East in the years after World War II. Another war, in Vietnam, would bring many of the ocean station cutters as well as Coast Guard patrol boats to that divided nation.

The Coast Guard first became involved in the Vietnam War in 1965 when twenty-six of its *Point*-class patrol boats (67 tons, 82 feet) were designated Coast Guard Squadron One and they and their crews deployed to South Vietnam to participate in an operation that began in March 1965 called Operation Market Time. The purpose of Market Time was the interdiction of arms-laden, unmarked trawlers that were steaming from ports in North Vietnam and negotiating the coastal waters of South Vietnam to spirit arms to the Vietcong, as well as the interdiction of junks and sampans transporting arms, troops, and supplies inshore. Market Time consisted of an outer screen (50 to 100 miles offshore) of patrol planes; an inner screen (40 miles offshore) of, initially, destroyers and minesweepers but later the *Edsall*-class destroyer escorts (which had been returned to active Navy duty after their work as ocean station cutters during the Korean War) and later still Coast Guard ocean station cutters; and inshore patrols by the Navy's 50-foot Swift boats and the Coast Guard's 82-footers.[1]

The shallow draft of the 82-foot patrol boats (less than six feet) made them ideal for work close inshore. Despite their light displacement and relatively thin skins, the patrol boats themselves suffered few casualties from combat. The worst damage inflicted on any of the 82-footers occurred when the U.S. Air Force mistakenly attacked the *Point Welcome* one night in August 1966, killing her commanding officer and one other crewman while seriously injuring five others.

The ocean station cutters and their crews were effectively brought into the war in 1967 as an augmenting force for the inner screen. In March 1967, the commandant of the Coast Guard, Adm. Willard J. Smith, directed that the East Coast cutters *Barataria, Half Moon*, and *Yakutat* as well as the West Coast cutters *Bering Strait* and *Gresham* depart their homeports to arrive at Pearl Harbor by late April and at Vietnam by early May. Thus, the ocean station cutters would wind up in Market Time alongside the Coast Guard's 82-foot patrol boats. The five *Casco*-class cutters and their crews, designated Coast Guard Squadron Three, left Pearl Harbor together on April 26 bound for Subic Bay in the Philippines.

While en route, the crews of the cutters repeatedly practiced evolutions that would be necessary in the war zone but that occurred only infrequently on ocean station: underway replenishment (receiving supplies from other ships), station-keeping ("flying in formation" for ships), and General Quarters (battle stations). Within five days of their arrival in Subic Bay on May 10, the cutters started patrolling the waters of Vietnam.

Coast Guard Squadron Three. From left to right: The cutters *Half Moon*, *Yakutat*, *Gresham*, *Barataria*, and *Bering Strait*.

The crews of the *Casco*-class cutters were really carrying out three missions at once: (1) interdiction of trawlers offshore inspecting them for weapons and war matériel; (2) fire support for forces ashore via use of the 5-inch guns; (3) provision of fuel, water, supplies, medical assistance, and engineering and electronics repairs to the much smaller patrol boats.

As the five cutters neared the end of their scheduled wartime deployment, two *Secretary*-class cutters, *Campbell* and *Duane*, along with three *Owasco*-class cutters, *Androscoggin, Minnetonka*, and *Winona*, prepared to relieve them. With more time for planning, these cutters had special shipyard periods, and their crews had supplemental training before sailing to Vietnam. Rather than have all five cutters arrive at the same time, the Coast Guard staggered the reliefs from December 1967 through January 1968 so that the "new" cutters could learn from the cutters already in place.

This system of ten-month deployments and staggered rotation would continue until 1970 when the number of cutters in Squadron Three was reduced from five to four. By January 31, 1972, the Coast Guard's contribution to Market Time came to an end. All of the Coast Guard 82-foot patrol boats had been turned over to South Vietnam in 1970 and were followed by *Bering Strait* and *Yakutat* in January 1971 and then *Castle Rock* and *Cook Inlet* in December of the same year. Then, in March 1972, the cutters *Absecon, Chincoteague*, and *McCulloch* were designated Coast Guard Squadron Two and dispatched together from Norfolk, Virginia, to Guam where they were decommissioned and transferred to the Navy, which in turn transferred them to South Vietnam. It was a rare event: Three ocean station cutters making a multithousand-mile transit all in company. Such a sight would never have been seen during the heyday of the ocean stations when the individual cutters simply went to and from station alone, just as they were alone while on station.

By the time the Vietnam War was over for the Coast Guard, thirty of the ocean station cutters, or 62 percent of the post–World War II ocean station fleet, had seen action in Operation Market Time.

A rare sight: The ocean station cutters of Coast Guard Squadron Two steam in formation.
Top to bottom: The cutters *Chincoteague*, *Absecon*, and *McCulloch*.

* * *

THE VIETNAM WAR was not the first time Coast Guard ocean station cutters and their crews had gone to war: The *Secretary*-class cutters had built an admirable record as escorts and in antisubmarine warfare during World War II. Nor was anyone shocked to learn of the assignment of the ocean station cutters to a combat mission. Warfare is part of the Coast Guard's duties. In fact, it was the need to maintain warfare readiness that sent each of the ocean station cutters to training with the Navy once every two years.

This refresher training (inevitably reduced to the acronym REFTRA) took the West Coast cutters either to Pearl Harbor or to San Diego and the East Coast cutters to Guantanamo Bay (inevitably reduced to the acronym GTMO) for drills, exercises, and training with the Navy. REFTRA usually lasted about a month and, except for the ships homeported in Honolulu, this meant another month away from home added to the months on ocean station.

There was more to REFTRA than gunnery, antisubmarine warfare, and fleet maneuvering. The ships also practiced towing and small-boat operations, though it was usually the Coast Guard crews with their frequent first-hand experience at these evolutions that "refreshed" the Navy instructors. A valuable part of the training was damage control procedures such as plugging leaks, pumping unwanted water overboard, and shoring up broken ship structures—the cutter crews had need of those skills while battling the weather on ocean station. And then there was firefighting training.

If a car catches fire, one can park it and get out. If a building catches fire, one can usually walk out to the street. If a ship catches fire, the only place to go is into the sea where one will face dehydration, exposure, starvation, and hostile creatures.

If a car or a building catches fire, someone calls the fire department, which will show up in a few minutes and battle the blaze. A shipboard fire has to be fought by the crew on the ship. It is not a question of saving the ship or saving yourself: You must save the ship *so that you may save yourself.*

A shipboard fire is unlike a shoreside fire in other ways. First, every possible source of fuel is present in huge quantities: hundreds of gallons of lubricating oil, thousands of gallons of fuel oil, paint on every surface, miles of wiring, dozens of bedsheets and blankets, and, on military ships, ammunition.

Second, water is the extinguishing agent that is most readily available and available in the largest quantity. Which is fine, except using water to extinguish a fire in the ship's interior means the firefighters are effectively putting the ocean inside their ship. They are, in other words, sinking their own ship.

Third, that water is saltwater. Saltwater is a wonderful conductor of electricity, which means that if it strikes an electrical object such as a light fixture or a radio or a radar set, it can short out the entire circuit. This means the firefighters may be quite literally in the dark.

They are probably in the dark anyway because of smoke. All fires produce huge clouds of smoke. In shoreside fires, the smoke almost always escapes because openings in the building allow it to rise and dissipate in the surrounding air. Aboard ships, that process has to be stopped because smoke is carried by heat, and if the heat is allowed to rise inside the ship, it will tend to start fires in spaces above the existing fire. Therefore, shipboard firefighters are generally taught to contain the fire within one space, cool the adjoining spaces, and then attempt to extinguish the fire.[2] This means the burning space is filled with smoke.

Real fires on real ships do not look like those in the movies. The firefighters do not see leaping flames. They cannot see the bulkheads, decks, machinery, or even each other. Often, the visibility is zero simply because the smoke is too thick. The only way the firefighters can find and extinguish the blaze is by looking for a glow in the otherwise uniformly gray-black atmosphere.

Firefighters on ships must work in teams. A lone person going into a smoke-filled compartment will be hard-pressed to keep track of his own location let alone find the fire, drag a hose along, and spray the blaze with it. Furthermore, the masks and breathing apparatus firefighters must wear limits their peripheral vision, so someone else, another member of the team, must keep looking over everyone's shoulder to make sure the fire has not sprung up behind them.

Because the heat rises, the overhead may catch fire as may the painted bulkheads and decks, trash cans, mattresses, or wiring. The firefighter can literally be surrounded by fire once inside the space.

Engine-room fires are particularly fearsome not only because the spaces are so big and filled with hazards, but also because a catastrophic engine-room fire means the ship loses propulsion and then becomes subject to the whims of the weather. An engine-room fire can also cause the loss of electrical power throughout the ship, which means not just darkness but also the absence of pumps used to pressurize the fire hoses.

Furthermore, in engine rooms, the decks on which crewmen walk are not the bottom of the space: they are just walkways built above the bilges. The bilges below the walkways are traps for oil and fuel leaks, and thus those fighting a fire in an engine room have to worry not just about the fire in front of them or above them or possibly behind them but also the potential fire beneath them.

Therefore, fighting a fire on a ship takes not just physical courage but also mental discipline and repetitive training so that fear is supplanted by learned reaction. That is what the firefighting training at REFTRA was all about.

THAT TRAINING PAID OFF more than once. In March 1968, the cutter *Barataria*, since returned from Vietnam, was sailing through the Unimak Pass while on a fisheries patrol in the Aleutian Islands. One of the cutter's diesel-powered generators started to smoke and run erratically. Engineman Third Class Jon T. Norstog hurried to the errant machine in an attempt to slow it down and adjust the speed control. Before he could complete his work, the generator erupted into flames. He grabbed a fire extinguisher and fought the fire. He had no breathing apparatus and no protective equipment.

As he fought the fire, the engine suddenly exploded; not violently or extensively, but it exploded nonetheless. The already existing fire had produced a good deal of smoke, and now the explosion added to the hot, dark cloud. If the engine was not stopped soon, it would continue to feed the fire and make even more smoke, which would force all the engine-room watchstanders out of the space. In the minutes it would take for the rest of the cutter's crew to man their firefighting stations, the engine room could fill with smoke and the fire could quickly rage out of control. One wrong move at this moment could doom the cutter. One right move could save it.

The explosion had burned Norstog's hands, arms, and head, but he stood his ground. The smoke was becoming dense, and he could barely see and hardly breathe. But he put out the fire and wrestled the engine to a stop. Catastrophe was averted. One of Norstog's shipmates groped through the smoke and helped the burned, choking, suffocating petty officer out of the engine room. His cutter would live and so would he. A grateful service awarded him the Coast Guard Medal for his "unwavering devotion to duty."

WHEN THERE IS NO ONE such as Norstog at the scene of an incipient fire, the results can quickly turn catastrophic. A smoldering ember can and will turn a room or a space on a ship into a blazing inferno in less than two minutes. If the tiniest fire goes unnoticed or unfought for even a few seconds, it may blossom into a raging inferno that simply cannot be extinguished. That is what happened aboard the passenger ship *Viking Princess* around midnight on April 7–8, 1966.

That night a fire, apparently caused by an exploding high-pressure air bottle, started aboard the *Viking Princess* while off the east coast of Cuba. Because almost everyone in the ship's crew as well as most of the passengers were asleep, the fire and smoke spread quickly, and the order was soon given to abandon ship.

After the radioman of *Viking Princess* made a distress call stating the ship was on fire and being abandoned, the Coast Guard sent an aircraft to the scene to determine the extent of the problem and to get more accurate information on the status of survivors. Meanwhile, the U.S. Navy responded by sending ships out of the Guantanamo training base to help. One of those ships was the Coast Guard cutter *Cook Inlet*, which was at GTMO for REFTRA.

Viking Princess' radioed position was wrong. The ship was really sixty miles south of where the radio message said she was. Regardless, it did not take the Coast Guard airplane or three nearby ships long to find her because she was fully ablaze in the dark night.

Viking Princess was lucky in that the weather was fair and three commercial ships, *Cap Norte*, *Navigator*, and *Chunking Victory*, were close by and began to recover her passengers and crew. But with almost 500 people divided among the three ships, a dark night, and the inevitable confusion in even an orderly abandonment, no one was sure if everyone had gotten off the blazing liner.

Cook Inlet arrived at the scene of the burning *Viking Princess* without knowing how many, if any, of the liner's crew or passengers were still aboard. With most or perhaps all of *Viking Princess*' crew already off, with no further radio communication to the burning ship, with no quick way of determining whether the people scattered among the other three ships constituted all hands, there was only one way to make sure no one was still alive and trapped aboard the cruise ship: Some of *Cook Inlet*'s crew would have to board the vessel and search it. Six of *Cook Inlet*'s crew quickly volunteered for the dangerous mission: Cdr. Ronald J. Dinsmore, Lt. Albert C. Buechler, Lt. (jg) Albert C. Davies, Chief Petty Officer John E. Johnson, and Damage Controlmen Third Class Robert S. Barnes and William O'Keefe.

These men went aboard an unfamiliar ship in the dark of night facing flames and smoke and uncertainty. They were not and could not be sure if their mission was even necessary. After all, everyone aboard *Viking Princess* might have already abandoned ship, in which case the Coast Guardsmen would be risking their lives for no reason at all. As the cruise ship's fire spread, it could engulf them. As they searched belowdecks on a ship they did not know, they could become lost in the smoke and suffocate as they tried to find their way out. As the fire burned, the ship's structures would eventually collapse, possibly crushing them or closing off their escape route. There were reports that the fire was so bad some of the ship's hull plating had already fractured, and thus the cruise ship might already be sinking or on the verge of capsize. There was almost infinite risk and only an infinitesimal chance that anyone was still alive. But the six men started their search.

They searched above and below including the infirmary. They worked quickly and got as close to the inferno as they dared. They kept searching, probing around the edges of the fire and advancing until they were sure no one could be left alive beyond the areas they saw. As they searched the fire got worse. The heat became so intense that paint on the outside of the ship's structure did not just melt, it burst into flames. They kept searching, braving the flames and the smoke until they realized that the fire was moving through the ship underneath them. They had fire in front of them and now fire beneath them. They had found no one and they could stay no longer. They returned to *Cook Inlet* and watched as *Viking Princess* turned into a blazing hulk.

In the end, eleven of *Viking Princess*' people were unaccounted for. Yet the six Coast Guardsmen made the maximum possible effort to save anyone who might have still been alive aboard that ship. All six were awarded the Coast Guard Medal for their "daring in spite of imminent personal danger."

* * *

THE ASSIGNMENT OF CUTTERS to Operation Market Time, the transfer to South Vietnam of the cutters *Bering Strait*, *Yakutat*, *Castle Rock*, and *Cook Inlet* in 1971 and of *Absecon*, *Chincoteague*, and *McCulloch* in 1972, along with other decommissionings and reassignments of cutters should have left a "hole" in the ocean station fleet; however, it did not. Though the fleet was being stretched thin, January 1972 would mark the beginning of the closures of the ocean stations, and the Coast Guard was already in the process of building an entire new class of offshore cutter before the transfers to South Vietnam started.

By the mid-1960s, it had been two decades since the Coast Guard had acquired a truly new, major cutter (the *Owasco* class) and three decades since the *Secretary*-class cutters were designed. The *Casco*-class cutters, which had first been borrowed then transferred outright from the Navy, were all approaching their twentieth anniversaries as well. The entire fleet of high-endurance cutters was aging and needed to be replaced.

Although the station duties were much the same as they had been in 1940, new technology permitted the design and construction of cutters better able to deal with the rigors of continuous offshore work. Machinery such as controllable-pitch propellers made ship-handling easier. Ships built with steel hulls and aluminum superstructures would not only be lighter in weight (and thus more efficient) but also have a lower percentage of weight topside, thereby increasing stability. Gas turbines, the same types of engines that powered airplanes, could provide small, comparatively lightweight, and extremely powerful propulsion units. (They were also extremely "thirsty," but fuel consumption was less important than speed when going on a search-and-rescue mission.)

As a result, the service designed and in 1965 ordered construction of a new class of high-endurance cutters (3,000 tons, 378 feet) to upgrade the aging "white-hull fleet." The first nine of the new cutters were named for secretaries of the Treasury with the very first cutter appropriately named *Hamilton*. The last three cutters were named for Coast Guard heroes: *Munro* (for Douglas A. Munro, the only Coast Guardsman ever awarded the Medal of Honor [posthumously] for his gallantry in the Pacific during World War II); *Jarvis* (for David A. Jarvis, who was awarded a special Gold Medal of Honor for a winter 1897 rescue in Alaska); and *Midgett* (for John A. Midgett, who was awarded the Gold Lifesaving Medal for rescuing the crew of a burning, sinking tanker during World War I).

If the *Owasco*-class cutters could only be described as ungainly, then the *Hamilton*-class cutters could only be described as svelte. They are sleek with narrow beams; long, raked, flaring bows; swept-back stacks; and gracefully angled superstructures.

Their engineering plants were designed to be versatile with two shafts driven by either diesels (for fuel economy and reliability) or gas turbines (for speeds of thirty knots during search-and-rescue work) and were designed with controllable-pitch propellers and a bow thruster for great maneuverability. The slow fuel consumption of diesel engines translates to long-term operation without refueling, a feature needed to patrol ocean stations.

These cutters were also designed with flight decks to accommodate the Coast Guard's helicopters, which had become an ever more important tool in search-and-rescue work. Because the cutters were built with the ocean stations in mind, they were constructed with a

Hamilton class: The cutter *Morgenthau*.

balloon shelter fitted between the two stacks and just forward of the flight deck.[3] All of the older ocean station cutters had these balloon shelters added on after their initial construction.

For all the advantages of the *Hamilton*-class cutters, however, they seemed to have one significant drawback as ocean station cutters: They rolled heavily in adverse weather. The cutters were designed with a lot of superstructure above the main deck compared with that of other ocean station cutters, and, even though designed much longer and heavier than any of the other ocean station cutters, they were designed with a beam only marginally wider than that of other ocean station cutters. When the Coast Guard cut down the stacks on these cutters to reduce topside weight, the cuttermen on the *Secretary*-, *Casco*-, and *Owasco*-class cutters cited that as "proof" that the *Hamilton*-class cutters were less able to take the adversity of ocean stations than were their own, older, smaller cutters.

THE NEW CUTTERS were not joining the fleet any too soon. Apart from the transfers to South Vietnam, decommissionings, and the assumption of new duties (such as oceanographic research and fisheries patrols), the older cutters were, in some cases, simply falling apart. The *Casco*-class cutter *Matagorda* was a case in point.

Matagorda's crew got an unwelcome Christmas present on Ocean Station Victor in 1964. On December 25, her crew heard unusual noises coming from one of the bearings supporting the starboard propeller shaft. Aboard *Matagorda* and all the other *Casco*-class cutters, the power from the engines drove reduction gears that were connected to a large, solid steel shaft. The other end of that shaft was attached to the propeller. Because the distance between the reduction gears and the propellers was long, the shafts had to be supported by a series of bearings mounted in the cutter's hull. It was one of these bearings that was making the noise.

Matagorda's crewmen found unusual rubbing on one of the oil seals. They found nothing else wrong, made repairs, and continued their patrol. Then, on January 11, 1965, the noise returned, and they made further checks. They found that the entire propeller shaft had shifted aft. Yet the connections between the engines and reduction gears and between the gears and the shafts were all normal. This was a real puzzle.

A puzzle until they discovered a number of breaks in the cutter's hull structure. There were broken frames and cracks in her decks. The combination of a shifted shaft and fractured frames might have been coincidental. But if not, these facts could lead to the conclusion that the entire cutter was bent. When and how this happened was anybody's guess.

Matagorda had been dry-docked four months earlier, and it was possible that the supports placed under her in dry dock had been improperly located permitting the cutter to twist while it was out of the water. Also, the cutter had been struck relatively hard by an assisting tug one day after the dry-docking, and that might have bent the hull.

But it was also possible that weather had beaten *Matagorda* out of shape: The cutter was a sister ship to *McCulloch* and *Half Moon*, both of which suffered structural damage in poor weather.

Her commanding officer, Capt. James L. Lathrop, reported his findings to the Fourteenth Coast Guard District not just because he was worried about his cutter and crew (he was), but also because at the time he was engaged in assisting a commercial ship, the *Santa Helena*, which was disabled near the northwest edge of Victor.

Matagorda had been relieved as Ocean Station Victor on January 9 by *Bering Strait* and headed directly for *Santa Helena* arriving at the stricken vessel's location that evening. Throughout the next two days, the wind blew out of the northwest at twenty-five to thirty-five knots, and the seas were running ten to twelve feet. Nevertheless, on the morning of January 11, *Matagorda* got close enough to *Santa Helena* to be able to shoot a single line (known as a "messenger" line) to her and thus pass over a radio. Then, *Matagorda*'s crewmen began discovering their cutter's problems.

Because of *Matagorda*'s cracks and questionable propeller shaft, *Bering Strait* was ordered to relieve her of escorting *Santa Helena* so that *Matagorda* could sail to the nearest port (Honolulu, which was also her homeport) where the damage could be evaluated and, everyone hoped, repaired. On January 13, *Bering Strait* arrived on scene, and *Matagorda* headed home.

Matagorda was instructed by the district office to "minimize working of the ship" and to reduce or eliminate "successive hogging or sagging cycling situations." No one at the district office, however, explained exactly how *Matagorda* was supposed to minimize sagging and hogging while making a week-long transit through twelve-foot seas. They only told Captain Lathrop that if he failed to do so, the main deck plating could start to crack and that such cracks "usually travel rapidly and serious damage can result."

Matagorda's crew, a typical Coast Guard crew, was not going to wait for someone else to solve their problems. As the crewmen headed home, they drilled holes at the ends of the cracks, a technique that will usually stop a crack from spreading. They added doubler plates where they could to reinforce the steel. They locked the starboard shaft so that it could not "freewheel" and thus further damage the shaft or bearings as they sailed home using only the port shaft.

After a week-long transit, *Matagorda* arrived in Honolulu where the Coast Guard's shore-based engineering staff began to evaluate the cutter's problems. It was obvious the cutter would have to be dry-docked again to make repairs. During the next two months, *Matagorda* was subjected to a number of tests, a dry-docking, and more than three weeks in a commercial shipyard as the Coast Guard worked to make the cutter seaworthy once more. Finally, in early April, her crew reloaded all the supplies, and *Matagorda* sailed once again for Ocean Station Victor.

The repairs were successful as *Matagorda* spent two more years making the long voyage from Honolulu to Victor and back again. But as the *Hamilton*-class cutter *Mellon* neared completion in October 1967, she was designated as a direct replacement for *Matagorda*, which was immediately taken out of service and decommissioned.

ENDNOTES

1. Uhlig (1994, p. 316). In his description of Operation Market Time, Uhlig notes that the 82-footer was better suited for the patrol than the 50-foot Swift boat.

2. Recently, this practice has come into question. Some shipboard firefighting schools now teach the practice of ventilating as a means of reducing heat within the burning space so that the fire will be less hot and can be fought more easily. All shipboard firefighting, however, is situational. There is no "textbook" answer.

3. Once the ocean station patrols ended, the balloon shelter was redundant. But the ever-innovative Coast Guard crews found good use for the space. As these cutters began sailing with embarked helicopters in the mid-1970s, the balloon shelter became a storage and work area for the aircraft. When the *Hamilton*-class cutters underwent a major modernization in the late 1980s and early 1990s, the Coast Guard "officially" did what the crews had already done: The balloon shelter was converted into a retractable hangar.

10

HOTEL

38°00' N, 71°00' W

HAVING BATTLED thirty-foot seas and sixty-knot winds for some twenty-four hours, the captain of the refrigerated ship *Bougainville* had his curiosity piqued when he sighted flashing red and white lights several miles from his ship on the evening of March 27, 1971. Swinging his ship around—no simple task in the gale that buffeted his and other ships off the coasts of Virginia and the Carolinas that evening—the captain of *Bougainville* closed the distance to the lights, hence to the radar target that seemed to be their source. As he approached, the lights ceased their intermittent nature and became steady lights, red over white. Not wishing to get too close to the vessel, which the radar picture showed to be large, the captain circled *Bougainville* around it at a safe distance, and *Bougainville*'s crew attempted to contact it by flashing signal light and by radio beginning around 1900, but to no avail.

The captain radioed the First Coast Guard District to report the sighting. After the Coast Guard duty officer checked with the U.S. Navy to ensure it had no operations in the area, the duty officer gave the captain the only reasonable answer: The lights and the object belonged to one of a group of fishing vessels known to be operating in the area. *Bougainville's* captain told the duty officer that what he was seeing through the darkness and spray "seems too big" to be a fishing vessel.

The captain concluded that perhaps there was nothing wrong after all. Besides, he had another ship on his radar only three miles away, and if anything had happened, the crew of the other vessel would have known about it. Finally, the seas were rough, maneuvering was an awkward and dangerous undertaking, and the first responsibility of *Bougainville's* captain was to the safety of his own ship and crew. He notified the Coast Guard that he would advise if anything changed and resumed his own course on a slow speed.

What the captain of *Bougainville* did not know, and what the Coast Guard duty officer did not know, was that *Bougainville* had just given a glimmer of hope to thirty-one desperate souls aboard the stern half of the tanker *Texaco Oklahoma*, which had split in two 15½ hours earlier about 100 miles southeast of Ocean Station Hotel.

* * *

STATION H HAD BEEN established at position 36°00' N, 70°00' W after the First Conference on North Atlantic Ocean Stations in 1946 and had been closed after the Fourth Conference on North Atlantic Ocean Stations in 1954. Yet on February 20, 1970, Station H was assumed at position 38°00' N, 71°00' W—less than two years before the Coast Guard began to close the other ocean stations.

Why was the station reestablished? The short answer is that Congress demanded it. The language in the Coast Guard's official documents states that "public opinion, as expressed through the Congress is demanding immediate solution" to a lack of warning about "catastrophic weather disturbances affect[ing] the eastern region of the United States."

Severe wintertime storms form in the Gulf of Mexico and/or in the waters off the North and South Carolina coasts and then move just offshore up the East Coast and through New England. These unnamed storms build quickly and deeply and move rapidly resulting in very little warning, violent winds, and often heavy precipitation, usually snow. They raise havoc not only with communications, transportation, schools, and businesses ashore but also with all those who make their living at sea. They are a genuine threat.

Thus, the reestablished Station H would serve to detect development of the type of storm system that could ravage the East Coast and that could break ships in two.

IF ANYONE HAD any lingering doubts about the potential severity of these storms, those doubts were eliminated 3½ months before a Coast Guard cutter first arrived on the reestablished Station H.

At 0941 on the morning of November 5, 1969, *Keo*,[1] a 16,000-ton, 628-foot tanker, reported its status as "cut in half, all crew on stern section" in more than fifty knots of wind with seas running fifteen to twenty feet. The Coast Guard's response was almost immediate and "by the book": The New York Rescue Coordination Center dispatched a C-130 airplane from Air Station Elizabeth City to locate the tanker and then serve as "cover" for a helicopter that would be dispatched from Air Station Brooklyn to evacuate the crew. The New York Rescue Coordination Center also directed the *Reliance*-class cutters *Vigorous* and *Decisive* (1,000 tons, 210 feet) to proceed to the scene. Then a second Coast Guard C-130 from Air Station Elizabeth City, a U.S. Air Force C-130 from Pease Air Force Base, and a Coast Guard helicopter from Air Station Salem were dispatched.

Less than an hour later and before any of the Coast Guard aircraft or cutters could arrive on scene, the crew of the yacht *Whisper*, a new Morgan 41 sloop, reported that the yacht was breaking up in a position about thirty miles southeast of *Keo*. As the Coast Guard's search-and-rescue case study would later note, both of these vessels were just north of a "small compact low pressure system" that was moving north. This storm was going to make the weather conditions surrounding both vessels far worse before it got better.

At 1050, *Rockaway* was directed to get under way, and *Rockaway* sailed from Boston one-half hour later. At 1232, the *Cherokee*-class cutter *Tamaroa* (1,600 tons, 205 feet) departed New York for the effort, and within thirty minutes *Campbell*, in Portland, Maine, was directed to make preparations to sail. Shortly after 1600, *Campbell* as well as *Yakutat,* the latter on Ocean Station Delta, were told to head for the scene.

Meanwhile, at 1140 the first Coast Guard C-130 arrived on scene. There was no sign of *Keo.* By noon, the other two C-130s were on scene. There was still no sign of *Keo.* One aircraft was dispatched to *Whisper*'s reported position. There was no sign of *Whisper.* No one in the search planes should have been surprised. Navigation being performed by people on vessels in storms who are preparing to abandon ship is often poor. The searchers all started looking for the halves of the tanker and the small sailboat.

Whisper was still afloat and still in one piece and still had a functioning radio, thus the aircrew of one of the Coast Guard C-130s asked the yacht's crew to make continuous broadcasts so that the C-130 could locate them through radio direction-finding. After three hours of searching, the C-130 aircrew found the yacht. The aircrew also managed to find a nearby tanker, the *Thomas*, and directed that ship to the sloop. By 2310, the tanker had managed to get alongside the now derelict sailboat and evacuate the entire crew.

Things were going far less well with *Keo.* The first sign appeared at 1700 in the late afternoon when the aircrew of a Coast Guard C-130 found a large oil slick and what appeared to be the partially submerged bow of a ship less than thirty miles from *Keo*'s last reported position. That none of this had been sighted in the preceding five hours is more commentary on the poor search conditions than on the competence of the searchers.

THE CEILING (i.e., the lowest layer of clouds) was at 300 feet. Hence, the airplanes had to fly below 300 feet to see anything on the surface. This is an uncomfortably low altitude, particularly over open water. Any failure in the aircraft's systems could quickly lead to an unintentional ditching. Consequently, the aircrew was paying a lot of attention to flying the airplane, which diverted some of them from actually searching.

One key to success in searching from the air for anything at sea is selecting the correct altitude. By getting as high as possible, one maximizes the distance one can see. However, flying too high will mean small objects are simply not seen. Conversely, flying too low can lead to the same result: A fixed-wing airplane will be flying at some speed well in excess of 100 knots, which means objects will be overflown quickly and thus may be missed. A helicopter, of course, can fly as slow as necessary to examine each and every wave if that is required. But helicopters lack the endurance to go, and stay, far offshore for an extended search.

Therefore, aircraft searches over water are typically conducted at altitudes between 500 and 1500 feet so that the airplane is low enough for the aircrew to see relatively small objects but high enough to look out over a great distance. A 300-foot altitude is simply too low for maximum effectiveness.

Poor visibility added to the problem of locating the bow section of *Keo*. The storm had rain, squalls, and fog, which limited visibility to two miles at the most and sometimes to zero. This meant the airplanes could have flown right over the remains of the ship, even several times, and still never have seen it.

It was discouraging to find *Keo*'s bow very close to the reported position after five hours of "boring holes in the sky," but at least the searchers knew they were looking in the right place. Confirming the location is critical to carrying out further searches. Once the location, or datum, is known, searchers will make special efforts to keep track of that position.

That position, however, moves. Whether it is a life raft or the bow of a tanker, any object in the water is subjected to wind and currents. It is extremely unlikely—practically impossible—for anything drifting in the water to stay in the same place. However, all objects in the water will react to the wind and current in a similar (though not identical) manner. Thus, the bow section would give searchers a good starting point for subsequent searches. If it stayed afloat.

TO GUARD AGAINST losing datum, Coast Guard aircraft were equipped with deployable radio transmitters known as datum marker buoys (DMBs). These rugged devices could be dropped into the water where they would remain afloat and send a radio signal out for several days. The C-130 proceeded to drop a DMB adjacent to *Keo*'s bow. This provided the aircraft and ships with a homing beacon and a point of origin for subsequent searches.

Searches throughout the night found nothing. The following morning, November 6, *Keo*'s bow section was relocated, and by 1040 *Rockaway* was on scene with it. Some thirty-five feet of the broken tanker's head were above the water, but no one was on the hulk or anywhere around it. Four Navy destroyers (*Forest Sherman*, *Fox*, *Fiske*, and *Leary*) joined the four Coast Guard cutters (*Rockaway*, *Decisive*, *Tamaroa*, and *Vigorous*) in the late morning and early afternoon searches for survivors. The crews of the searching vessels found a lot of debris, oil, an empty life raft, and one of *Keo*'s empty lifeboats. The crew of USS *Fox* found a body.

Thus, the searchers now had further confirmation that they were in the right location. They also had confirmation that, in at least one case, they were too late.

They searched all night. Nothing.

On November 7, the vessels continued searching and found more debris and oil. At 1920 that evening, the crew of USS *Fox* found a man in a life jacket. The man was dead and his body mutilated by sharks.

They searched all the rest of the night. Nothing.

On November 8 at 1345, the crew of *Yakutat* found two oil-covered life jackets that had been joined together. Shortly thereafter, the cutter crew found four men in oil-saturated life jackets. All four were dead. Most of the bodies were badly mutilated.

That evening at 1800, the crew of *Tamaroa* watched *Keo*'s bow sink.

The search continued for one more day. The searchers found a few more things: a life ring from *Keo*, an unopened life raft still in its storage case, and some more life jackets including one that was properly made up for wearing and still had a shirt, a pen, and a religious medal in it. That life jacket had obviously been used by someone; that person was now gone. The crew of *Vigorous* found another life jacket with a body in it; the legs were missing.

During the course of the four-day search, there had been no signs of survivors and too many signs that anyone who had abandoned ship successfully was no longer alive—the oil, the exposure, and the sharks had got them all.

The Coast Guard terminated the search late on the afternoon of November 9. No one from *Keo* had been recovered alive.

LOSSES SUCH AS THAT of *Keo* and her crew demonstrated the need to reestablish Station H.

Ocean Station Hotel was located about 200 miles off the East Coast between Cape Cod and Cape Hatteras. Hotel was distinguished from the other ocean stations by its sole purpose as a weather station and by its relative proximity to a coastline. The grid was not used on the station; rather, cutters on Hotel were instructed to remain within a 15-mile radius of the station position.

Ocean Station Hotel was further distinguished from the other ocean stations by its operations schedule. The station did not have to be manned all year, only from August 1 to March 31. The dates were selected to cover the nor'easter season, though the period also included the height of hurricane season. The "concept of operations," however, additionally called for a cutter in standby to assume Hotel on 24-hour notice throughout June and July (the beginning of hurricane season).

A special weather radar was provided by the National Weather Service (formerly the U.S. Weather Bureau) for use on Hotel. This was a large, expensive, and permanent installation, and there would not be two such radars.

Therefore, a single cutter with the single weather radar installed would assume Ocean Station Hotel almost continuously for the 8-month nor'easter/hurricane season. After the 8-month season, the cutter would be on 24-hour standby throughout June and July of each year.

TO BEGIN OPERATIONS on Hotel, it was first necessary to select a cutter to assume the station.

The new station needed to be manned at a time when the ocean station fleet was already stretched thin by decommissionings and by service in Vietnam. Thus the Coast Guard looked for an "interim solution," and that interim solution would be the *Casco*-class cutter *Gresham*.

Gresham had already been taken out of the ocean station schedule in preparation for her decommissioning. Therefore, assigning her to the duty could provide a comparatively quick manning of the station with the least disruption to the rest of the cutters' schedules.

Gresham's homeport was in Alameda, California. If the cutter was to become Ocean Station Hotel, then she and her crew would all have to relocate to the Norfolk, Virginia, area, the port closest to the reestablished station. In the meantime, *Gresham*'s decommissioning process had been proceeding toward an October 30 decommissioning date.

TAKING A MILITARY SHIP out of service is not a case of just mooring it somewhere and disbanding the crew. A number of valuable items must be dealt with properly. Ammunition must be off-loaded and accounted for. This alone is no simple task as these cutters carried hundreds of rounds of powder and projectiles for the 5-inch gun as well as for the 81-mm mortars, 50-caliber machine guns, and other small arms. All that material had to be correctly counted and turned over to a weapons facility.

Similarly, classified material must be accounted for and properly destroyed or turned over to an appropriate unit.

Tens of thousands of gallons of fuel and lubricating oil were carried in these cutters. Those liquids had been paid for by the Coast Guard, and the service was not going to just give it to whoever took over the vessel. There had to be a procedure by which the "fuel and lubes" were transferred to other Coast Guard units.

Damage-control equipment was also valuable and could be used by other cutter crews. Breathing apparatus, fire hoses, pumps, eductors, pipe plugs, and firefighting foam were all expensive and useful. It took time to figure out who most needed this now-surplus gear and then money and effort to get it transferred there.

The list went on: Commissary supplies (groceries) had to be used up or transferred. Life rafts and ship's boats could be used by other cutters. Electronic equipment such as loran receivers, radios, and radar could either be transferred to units with older models or simply taken ashore and saved as spare parts. Other navigation equipment was also valuable and still had life left in it: sextants, bearing circles, alidades, plotting instruments, and binoculars. Bolts, books, bed sheets, blankets, coffee mugs, cooking utensils, cutlery, dishes, filters, flags, foul-weather clothing, hoses, life jackets, mooring lines, paint, paper, pencils, screws, typewriters (it was 1969), and tools of every description might have been worth little on an individual basis, but when such supplies number in the dozens or hundreds, they become valuable.

Finally, but of the greatest importance, is taking care of the crew. Some may be eligible for early discharge from the service rather than incur the time and expense of a short-term transfer, but that processing does take time. All the rest have to be reassigned someplace. For those with families, it may mean packing up and moving cross-country or overseas, something that cannot be done quickly or inexpensively. For everyone it means finding a new assignment appropriate to an individual's skills, experience, and desires. This is a time-consuming process that must be handled carefully: these assignments affect people's lives.

Thus there is a significant amount of planning involved in decommissioning, and everything cannot wait until the last minute. In *Gresham*'s case, she was scheduled to be decommissioned on October 30, 1969, and so the processes started in early September. Then, on September 30, the commander of the Twelfth Coast Guard District ordered *Gresham* to "Hold all orders in abeyance." The cutter's fate was being decided at Coast Guard Headquarters in Washington, D.C.

The original decommissioning date came and went. Most of *Gresham's* crew had already been issued orders and were expecting to transfer. Some of those expectations were put on hold. The removal of equipment was stopped. Plans for sailing the cutter to the East Coast, which plans had not existed thirty days earlier, were begun.

The decision to assign *Gresham* to assume Ocean Station Hotel was not finalized until November 1969. In early January 1970, Coast Guard Headquarters announced officially that *Gresham* would become the primary cutter on Hotel. The personnel offices began issuing new orders for the cutter's crew. Most would stay aboard long enough to get the cutter moved and to complete the first patrol on Hotel while a few were to remain aboard *Gresham* on a permanent basis. Thereafter, the bulk of *Gresham's* crew would be reassigned, and new men would be sent in to take their places.

Although the cutter would get an almost entirely new crew before summer, the size of that crew would be the same as before. However, because *Gresham* would no longer have any military readiness requirement, this effectively left "extra" men aboard; that is, there was no need for gunner's mates or sonar specialists, so their billets were filled with extra cooks or seamen or engineers or bridge watchstanders. Thus, according to the plan, *Gresham* could always leave part of the crew behind thus helping to alleviate the burden of the patrol schedule.

On January 15, 1970, *Gresham* sailed for Norfolk via the Panama Canal arriving at her new homeport on February 7. After a brief shipyard period, *Gresham* assumed Ocean Station Hotel on February 20.[2]

Gresham would slug it out with the North Atlantic through the end of the first (1969–1970), partial season on Hotel and throughout the subsequent three seasons. The special weather radar was installed and enclosed in what appeared to be a giant fiberglass golf ball above the cutter's wheelhouse, and the cutter's 5-inch gun was removed. The East Coast ocean station cutter schedule was changed so that different cutters could periodically relieve *Gresham* for brief periods, enabling *Gresham* to return to Norfolk for supplies and some crew change. But most of the time, from August through March, *Gresham* would be on Hotel.

* * *

SO IT WAS that *Gresham* was finishing up the 1970–1971 season on Hotel when *Texaco Oklahoma* broke in two about 100 miles southeast of the station.

Texaco Oklahoma had been built in 1958 as the tanker *Atlantis* at the Sparrows Point, Maryland, shipyard of the Bethlehem Steel Corporation. In December 1960, she was renamed when Texaco began an eighteen-year charter for her. She was typical of her era: 20,084 tons, 660 feet in length, with a 15,000-horsepower steam turbine propulsion system and thirty cargo tanks, which would bring her full-load draft down to about thirty-five feet. *Texaco Oklahoma*, again typically, had bridge and deck personnel accommodations amidships, whereas the engineering spaces and accommodations were located aft. There were four lifeboats aboard, two at the midships superstructure and two near the aft accommodations. Each of these boats was rated for 37 people, so she had more than enough boats to accommodate the entire crew of forty-four.

Texaco Oklahoma had consistently and satisfactorily completed all her required American Bureau of Shipping (ABS) and Coast Guard inspections, the most recent ones, including a dry-dock examination, being finished in July 1970. Nevertheless, the ship was scheduled for a biennial inspection in summer 1971. Coast Guard regulations at the time allowed one of the ship's boats to be stripped and cleaned in preparation for the inspection even though the ship was under way. The overabundance of boats onboard made this a reasonable practice. Consequently, *Texaco Oklahoma*'s port-side aft lifeboat was effectively out of service as she sailed from Port Arthur, Texas, to Boston, Massachusetts, on the afternoon of March 22, 1971, with a cargo of 220,000 barrels of black oil.

By March 26, the National Weather Service was reporting a gale-force low-pressure center off the coast of South Carolina and forecasting northwesterly winds of thirty to forty knots (with higher gusts) with accompanying rain and snow. Coastal stations were reporting seas of ten to twenty feet. *Texaco Oklahoma* was in the middle of this forecast area.

As the winds whipped the sea into a frenzy, *Texaco Oklahoma*'s speed was reduced from her full speed of about 93 propeller-shaft RPMs (about 15 knots) to 86 RPMs and then to 75 RPMs. She was still headed in the right direction (i.e., toward Boston), but the increasing size of the seas necessitated that she slow down to avoid having too much water break over the forecastle. By 2015 on March 26, the ship's speed was once again reduced, this time to 65 RPMs. *Texaco Oklahoma* was riding out the storm, but only just. Seas were washing across the decks and swirling around the piping and superstructures before spilling overboard. The vessel began to roll as well as pitch into the building seas. Wind howled around the bridge wings, and the ship shuddered as it pounded through the inhospitable North Atlantic seas. The captain reduced speed to 60 RPMs.

Rather than experiencing the northwesterly winds of thirty to forty knots that were forecast, *Texaco Oklahoma* was now in the midst of sixty-knot winds out of the northeast and forty-foot seas, which made further transit in the direction of Boston nearly impossible: The ship had to head into the seas to remain afloat until the weather abated. Progress became secondary to survival. By midnight, speed was reduced to 50 RPMs, which gave her about five knots in calm water and near zero knots in the howling night.

At 0330 on March 27, *Texaco Oklahoma* took a combined pitch and roll to starboard that, while severe, would not have seemed unusual except for one thing: When the pitch and roll stopped, the crew heard a loud cracking noise followed by a series of thumps. The men who were up and about on the stern section realized that the ship did not appear to be rising to the next sea but instead seemed to be staying down by the bow. They roused their shipmates in the aft accommodations.

When most of the crew came on deck, they peered through the cold driving spray to find an almost unbelievable sight: The forward half of *Texaco Oklahoma*, including the midships deckhouse and bridge, was no longer attached to the half on which they were standing. Cocked upward at an angle of some twenty degrees, the bow was above the waves, and the deckhouse, where thirteen members of the crew including the captain were either asleep or on watch, was awash in the monstrous seas that swept over the shattered ship.

None of the people who were quartered in *Texaco Oklahoma*'s midships deckhouse would live to tell of the catastrophe. The 50°F water that rushed into the forward berthing spaces was clinically and psychologically shocking. Most of those men likely lost consciousness quickly and drowned without ever fully understanding what happened to their ship.

Apparently, at least one crewman did not meet such a swift fate. The men standing on the stern section saw a light flashing in the wheelhouse. It seemed to the survivors that someone was attempting to signal from within the ship's bridge.

Stunned by the sudden breakup of the ship and immersed in total darkness because the cables that carried electrical power from the ship's generators aft were now severed, this crewman was likely lurching about searching for a life jacket and a useable door or maybe an open porthole through which to escape.

The source of that light and the thoughts and actions of the sailor wielding it shall never be known. That light was the last trace of life ever seen on the forward half of *Texaco Oklahoma*.

In the darkness and with seas crashing all about them and the wind tearing at their bodies, the men on the stern section of *Texaco Oklahoma* saw the bow section, like some angry leviathan, begin to drift down upon them. It struck and began to grind against the starboard side of the stern section. The bow-up attitude of the ship's forward half placed the ship's anchor at the same height as the lifeboats on the stern section. The anchor proceeded to demolish the starboard lifeboat, scattering its pieces into the sea. The port lifeboat was useless because it was stripped for the inspection, and now the starboard boat was gone.

The bow continued to grind against the stern. Knowing that their only hope for survival was to separate the two halves of the ship before the grinding either holed the stern section or created a spark that would start a catastrophic fire, the engineers backed the ship's propulsion, carrying the stern away from the bow.

Texaco Oklahoma's crewmen considered their circumstances: Their ship was broken in two, their captain probably dead (and if not, still on the bow, which had disappeared into the screaming night), and the one fully equipped lifeboat on the stern was destroyed. With no captain, only half of a ship, and a still raging storm, one choice was to make preparations to abandon.

However, the stern appeared stable; it had no list and it was not sinking appreciably even though it was down by its forward end. They had light and power because all the machinery spaces were located in the stern section. They decided to wait, knowing that it was more likely someone would see half of a ship than see a small life raft.

When dawn came on the morning of March 27, *Texaco Oklahoma's* bow appeared once again, threatening the stern. There was no sign of life on the ship's forward half, and, once again, the crew backed the stern away from the danger.

Throughout the day the crewmen on the stern section saw other ships, and they signaled the other ships using flares and the ship's whistle, but no one saw them—except for the captain of *Bougainville*, who made signals and radio calls at 1900 to the unknown object that had flashing then steady red and white lights.

The stern section of *Texaco Oklahoma* developed a starboard list and began to go down by the head.

BY 0330 ON MARCH 28, the stern section was down thirty degrees by its forward end. The crewmen fabricated two rafts from oil drums and made their one inflatable raft ready to launch. As the crewman faced their second dawn on the hulk, the bow-down angle increased to fifty degrees. It was time to leave.

At 0530, the thirty-one men on the stern section, all of them wearing life jackets, began climbing down a ladder to board the inflatable raft or the two oil-drum rafts. Their planned abandonment was going as well as it could. Fourteen or fifteen men got into the inflatable raft, and the remaining men got onto one or other of the oil-drum rafts. Then two things happened at once: The line holding the inflatable raft to the stern section parted, casting the raft adrift, and one of the cargo tanks ruptured, sending forth a wave of oil that swept over all the rafts and dumped everyone into the sea. Eleven men struggled back into the oil-covered inflatable raft. The rest of the crew grabbed at flotsam, life rings, and the slickened oil-drum rafts, but they all drifted apart. An hour after the crew abandoned the stern section, those who were still alive saw it sink.

The fabricated oil-drum rafts capsized repeatedly making them next-to-useless. The passage of time and fatigue, huge seas, darkness, and oil took their tolls. Those in the water began to drift off, unable to hold onto the makeshift rafts or each other. By the time *Texaco Oklahoma* was scheduled to arrive in Boston, the eleven men on the inflatable raft could only see each other.

When the tanker failed to appear, the Boston pilot called Texaco's representative. That started a series of phone calls from one Texaco office to another and from Texaco to the Coast Guard. Finally, at 1500 on March 28, Texaco asked the Coast Guard to begin a search for the ship.

Two hours later, the crew of the tanker *Sasstown* sighted a red raft adrift off the coast of Virginia. Approaching the raft and sounding her whistle to draw the attention of the people in the raft, *Sasstown* finally got close enough to find eleven men in the tiny craft. Despite the weather, *Sasstown* got alongside the rubber vessel, and her crew brought the men aboard. The crew then radioed the Coast Guard that the tanker *Texaco Oklahoma* had broken up the day before. The discovery of those men would be the first definite indication received by the Coast Guard that *Texaco Oklahoma* was gone.

The Coast Guard's Rescue Coordination Center in Norfolk immediately dispatched a C-130 airplane from Air Station Elizabeth City. The aircrew of the C-130 was to do several things: (1) establish direct voice communications with *Sasstown* to get accurate, up-to-date information on *Texaco Oklahoma*; (2) begin searching for other survivors, and, if any were located, (3) drop supply-laden rafts to them; (4) act as an on-scene coordinator for the search effort; and (5) obtain accurate on-scene weather so that search planning could be accurate.

WHILE THE FIRST Coast Guard C-130 started searching for *Texaco Oklahoma* survivors, the Coast Guard's radio station in Portsmouth, Virginia, began broadcasting a distress message to other ships in the area. The cutter *Escanaba* and eleven other commercial ships responded and headed for the scene.

The first vessel to arrive was the *Texaco Nebraska.* She went directly to the position where *Sasstown* had recovered the eleven men and began a search. *Sasstown* departed heading for her New York destination. *Texaco Nebraska* had little chance of finding anyone in the dark of night and fifteen- to twenty-foot seas. The only real hope was that any remaining survivors had lights or flares or that, perhaps, the bow section of *Texaco Oklahoma* was still afloat. If it was, no one would see it.

By noon the following day, March 29, *Texaco Illinois* and *Texaco Wyoming* had joined the search, and the *Cherokee*-class cutter *Chilula* was dispatched from Fort Macon, North Carolina, at 1210. Two hours later, *Texaco Nebraska* found two survivors from the stern section floating in their life jackets. She also saw an empty life raft, and a Coast Guard C-130 found two oil drums lashed together about five miles away. The searchers knew they were in the right location.

The cutters *Gresham* and *Rockaway* were directed to join the search along with a U.S. Air Force C-130 and the U.S. Navy's USS *Concord* with two embarked helicopters. As night fell, the crew of *Escanaba* retrieved an eighteen-foot-long section of broken brow (gangway) and tried to recover three bodies, which had no life jackets. The crew was unable to do so in the dark and lumpy sea. All rescuers would search throughout the night but find nothing else.

The next morning, March 30, saw the four cutters, the USS *Concord* and her two helicopters, the Texaco tankers *Nebraska, Illinois, Wyoming, Florida,* and *New Jersey,* the tanker *Thomas M.*, three Coast Guard C-130s, one Coast Guard flying boat, and two Air Force C-130s all searching for survivors. They covered more than 13,000 square miles of ocean. No survivors were found.

On March 31, the same ships plus *Texaco Rhode Island* and *Texaco Georgia* joined with four Coast Guard and two Air Force airplanes to search another 12,000 square miles. No survivors were found.

The tankers were all released from the search that evening, but the Coast Guard cutters, Navy ship, and Air Force and Coast Guard airplanes started again on April 1. They found three empty life jackets and nothing else.

They went at it again on April 2, and *Gresham* found a life jacket stenciled with the words "Texaco Oklahoma." The life jacket was tied up as if it had once been occupied by a human being. There was some other debris and an oil slick, but no survivors.

After one more day's effort on April 3, the Coast Guard believed the residue that had been found (life jackets, debris, oil) proved that the search had been in the right areas. More than 50,000 square miles of ocean had been searched or researched. No one was found alive after *Texaco Nebraska*'s recovery of two survivors on March 29. It was practically impossible that anyone could still be alive after six days in the open ocean with no food, shelter, or fresh water.

The Coast Guard called off the search.

THE END OF the search for *Texaco Oklahoma* in early April meant the end of *Gresham*'s 1970–1971 season on Hotel. *Gresham*'s crew would suffer another season in 1971–1972 and yet one more in 1972–1973.

But *Gresham* was near her end. The cutter had already been put on the chopping block in 1969, a clear indication that her service life was all but over. Three full seasons on Hotel battling the same type of weather that destroyed *Keo* and *Texaco Oklahoma* had done nothing to increase her longevity. The Coast Guard decided, again, to decommission the cutter. This time, the plans would not be changed.

The Coast Guard had to find another cutter to fill the billet. The special radar unit would have to be installed on that cutter, and that cutter and her crew would have to face the same tough schedule. Logic dictated using a cutter already homeported in the Hampton Roads, Virginia, area to avoid the personnel transfers and logistics challenges of moving a cutter from somewhere else. Logic prevailed, and the venerable *Taney*, in service since 1936, was selected to be the new principal ship for Ocean Station Hotel.

Taney had a secure place in Coast Guard history no matter where she was assigned because of what her crew did on December 7, 1941. On that day, *Taney* was moored in her homeport of Honolulu. As Japanese planes bombed and strafed Pearl Harbor and the surrounding area, *Taney*'s crew responded with antiaircraft fire. *Taney* would become known throughout the Coast Guard as the "ship that fought at Pearl Harbor" even though she really fought at Honolulu, several miles away.

During the years of World War II, *Taney* was in the Pacific until late 1943, during which time the cutter escorted merchant ships, evacuated U.S. citizens from far-flung islands, conducted antisubmarine patrols, and searched for survivors after the Battle of Midway. *Taney* then steamed to Boston to join the rest of the Atlantic convoy escort fleet.[3]

En route, *Taney* put in at Mare Island, California, to receive additional 5-inch guns, becoming the only *Secretary*-class cutter to have four of those weapons. She looked like what she was—a war-fighting ship.

Taney arrived at Boston on March 14, 1944, then proceeded to Hampton Roads, arriving on March 31. *Taney* conducted a series of convoy escorts until late 1944. The cutter then returned to the Boston Navy Yard where she was outfitted as an amphibious command ship, a role she would keep until the end of the war. *Taney* was then sent back to the Pacific, making port at Pearl Harbor on February 22, 1945, and subsequently arriving for combat operations off Okinawa on April 11.

After World War II, *Taney* would arrive at Charleston, South Carolina, on November 29, 1945, to be refit, then was assigned to Alameda, California, from where her crew patrolled the Pacific ocean stations until joining Coast Guard Squadron Three in Vietnam from May 1969 to January 1970. *Taney* returned home to Alameda in early 1970 and went back on Pacific ocean station duty for two years. Then, to rebalance the location of the cutter fleet caused by continuing cutter decommissionings and the closure of Ocean Station Victor, *Taney* was reassigned to Norfolk in February 1972. Thus, the cutter was in the right place at the right time to take over *Gresham*'s duties on Hotel.

Consequently, *Gresham's* decommissioning would be completed on May 1, 1973, after the end of the season and after the special radar was removed. *Taney* was ordered to have the radar installed and became "the major ship resource in the ocean weather station Hotel project" beginning in the 1973–1974 season.

There was one difference, however, between *Taney's* and *Gresham's* assignments. *Taney* was to "retain multi-mission capability." Her weapons, including the 5-inch gun, would remain onboard as would all the personnel filling billets related to warfare capability. And there would be no augmenting crewmembers to permit a percentage of the crew to remain ashore during the long, dark winter months. To compensate for this, the Coast Guard would reduce the amount of time *Taney* spent on station (compared with that for *Gresham*) by having other cutters spend more time on the station.

Taney fitted out with special radar dome on Ocean Station Hotel.

This was possible because the other ocean stations were being closed: Victor on January 15, 1972, and Echo on June 30, 1973. Delta was closed on June 30, 1973, and Charlie on December 31, 1973. Bravo was closed on June 30, 1974, and November was closed on June 30, 1974.[4]

Thus, the 1974–1975, 1975–1976, and 1976–1977 seasons on Hotel would mark the end of the U.S. Coast Guard's ocean station patrols.[5]

TANEY'S DEPARTURE FROM Ocean Station Hotel on the last day of the 1976–1977 season was the last time any Coast Guard cutter would leave any ocean station. The patrol was simply stopped without a celebration or observance, though there were few tears shed. The occasional days of glory on or en route to station were far outnumbered by the days of boredom and of battering by relentless seas. Many on active duty in the Coast Guard in 1977 had never been on an ocean station. Those who had did not miss it or want to go back.

ENDNOTES

1. *Keo* was one of only five vessels of the ill-fated *Bulkpetrol* class of tankers, all of which were built between 1948 and 1950 by Welding Shipyards of Norfolk, Virginia, a yard that had built T2 tankers during World War II. Of the five tankers, only the namesake of the class, the *Bulkpetrol*, survived. All had been renamed. The tankers *Keo*, *Pacocean*, and *Amphialos* each broke in two while under way during storms. The final tanker of the class, *Golden Drake*, exploded and then sank while under tow. (Source: Whittaker, A. Davis, Jr. "The *Bulkpetrol* Class Ships." The T2 Tanker Page. 20 February 2005. Available at: http://www.t2tanker.org/ships/bulktrader.html. Accessed 1 June 2009.)

2. It could not be determined from available Coast Guard historical sources if the National Weather Service radar unit was ready and/or installed on *Gresham* and the 5-inch gun mount removed during the brief shipyard period occurring between February 7, 1970, and February 20, 1970, or if *Gresham* assumed Hotel without the unit per the congressional requirement to begin operations before the end of the 1969–1970 season. A few weeks after end-

ing the season on March 31, 1970, *Gresham* entered a six-week shipyard period, and the work may have been done at that time.

3. Details provided in the sketch of the service life of Taney are taken from: "*Taney*, 1936: WPG / WAGC / WHEC-37, ex-*Roger B. Taney*." U.S. Coast Guard Historian's Office, Washington, D.C. Available at: http://www.uscg.mil/History/webcutters/Taney_1936.asp. Accessed 1 June 2009.

4. U.S. Coast Guard Commander, Twelfth District (1972); U.S. Coast Guard Cutter *Campbell* (1973); U.S. Coast Guard Headquarters (1973); Shuhy (1974, p. 4, chart); U.S. Coast Guard Commander, Fourteenth District (1974); U.S. Coast Guard Headquarters (1974).

5. White (1977); Siler (1978).
Ocean Station Hotel was formally closed on September 30, 1977.

11

ZULU ZULU

SOMETIMES A CREW on an ocean station would need to leave the grid for search-and-rescue operations or would be driven off the grid because of heavy weather. Sometimes a crew would become navigationally lost: Overcast skies prevented celestial navigation and atmospheric conditions disrupted loran transmissions or the weather damaged the loran receiving antenna.

When a crew knew that the cutter was not inside the grid or did not know where in the grid the cutter was, the position group that was broadcast by the cutter's radiobeacon would be changed to ZZ (Zulu Zulu).

This broadcast was a notice: The small ship in the middle of the ocean was not where it should be. It was sometimes also a signal: The cuttermen had reached a limit of man and machine.

* * *

ON MARCH 27, 2007, a U.S. Coast Guard helicopter successfully hoisted an injured crewman from the *CSCL Genoa*, a merchant ship that was located about 430 miles southwest of Bermuda. The helicop-

Pontchartrain and crew in heavy seas on Ocean Station Bravo (undated photo).

ter had departed Air Station Elizabeth City, North Carolina, in the late afternoon and was escorted by a Coast Guard C-130 airplane from the same base. Between Elizabeth City and the *CSCL Genoa*, the helicopter landed aboard the *Arleigh Burke*–class destroyer USS *Oscar Austin* for refueling. The helicopter then completed the flight to the *CSCL Genoa* arriving well after dark, hoisted the man, and delivered him to emergency medical personnel at Wilmington, North Carolina, just before 2300 that evening.

This case is typical for the Coast Guard. What makes it relevant here is that this "routine" medical evacuation contains all the elements that made the ocean stations obsolete in the 1970s: (1) reliable, all-weather navigation systems; (2) meteorologic forecasts so accurate that the comparatively fragile helicopter could make the flight with a near-guarantee that weather would not impede the operation; (3) aircraft, both fixed-wing and rotary-wing, with engines and flight control systems so reliable that the aircraft could operate hundreds of miles offshore with a very high probability of return; (4) rescue aircraft that could complete an evacuation far offshore and return safely without being exposed to the hazard and uncertainty of water landings.

NAVIGATION SYSTEMS. Ship navigation changed little in the years between the development of a reliable, mobile chronometer in 1761 and the onset of World War II. Seamen used an accurate timekeeping device and a sextant to obtain a navigational fix not because it was the best technology but because it was the *only* technology.

But that technology is not reliable. It is awkward, though not impossible, to take celestial observations (sun, planets, stars) if a ship is rolling or pitching badly. Such awkward observations will oftentimes be poor resulting in a poor fix. If it is overcast, those observations are impossible.

Radio navigation was the technological breakthrough that made reliable, all-weather offshore navigation a reality. Whereas the first radio-navigation systems enabled ships and aircraft to home on radio-beacon signals,[1] loran would preferentially provide navigators with an actual navigational fix and had a greater service area offshore. The early mode, Loran, was replaced by Loran-A, which was replaced by

Loran-C, which was supplanted by the ubiquitous GPS. Simultaneous with the development of electronic navigation, onboard inertial navigation systems—originally developed for rockets and missiles—were applied to aircraft navigation and refined, and extreme navigational accuracy became the norm.

Once aircraft were fitted out with good navigation systems, aircrews could not easily become lost over the oceans.

WEATHER FORECASTING. Because the ocean stations were few in number and located in only two oceans, the synoptic reports of the stations collectively did not provide information about weather conditions on either an oceanic or a global scale.

The ability to obtain a global assessment of weather conditions began in 1959–1960 with the advent of weather satellites. The first weather satellites provided pictures of cloud cover. As technology advanced, satellites could ascertain temperature, pressure, and humidity and could gather such data over large areas of the earth.

In the 1960s, the determination of surface wind speeds and wave heights began to be accomplished by buoys equipped with instrument packages and radio transmitters. Anemometers and vertical displacement devices on buoys collected data on surface wind speeds and wave heights, respectively, and the data was transmitted to forecasters ashore. Eventually, the United States would deploy dozens of such buoys off the nation's coasts.[2]

Consequently, in the 1970s, weather forecasters could get most of the weather data they needed by satellite and buoy. The forecasters had a broader and more timely overall picture of conditions and thus could make more accurate and far-reaching (both in distance and time) forecasts without depending on ships at sea.[3]

RELIABILITY OF AIRCRAFT. The jet engine was developed during World War II and was first installed in military aircraft. It would be more than a decade before jet engines were considered sufficiently reliable to be safe for use in commercial airplanes. Ironically, jet engines would be so reliable that they would make "conventional" (i.e., piston-powered engine) airplanes effectively obsolete within another decade.

A piston-powered engine is an inherently complex device. There are hundreds of interrelated metal parts that spin, slide, change direction, compress, expand, and rotate thousands of times each minute. If any one of these parts fails, it can and often does lead to the failure of the entire engine.

In an automobile, such engine failure only means the vehicle and its occupants will come to an unexpected stop. In an airplane, it means the aircraft will have to land no matter where it is unless it has enough supplemental power (i.e., additional engines) to stay airborne. It was the very probability of such failure in piston-powered airplane engines that mandated multiple engines in transoceanic commercial aircraft.

Compared with a piston-powered engine, a jet (or turbine) has very few moving parts, and nearly all the motion is rotational. This reduced complexity means greater efficiency as well as greater reliability. It creates greater power for a given weight. A jet, or turbine-powered, airplane is simply faster than a piston-powered airplane. Thus, allowing for headwinds, a trans-Atlantic flight in a jet will take about four hours, whereas that same trans-Atlantic flight in the flying boat *Bermuda Sky Queen* would take about seventeen hours.

Higher speed can also mean the difference between completing and not completing a flight. In 1947, the *Bermuda Sky Queen*, capable of an airspeed of 130 knots, had to ditch on its 1,730-mile flight because overloading and strong headwinds caused the aircraft to burn more fuel than anticipated and greatly affected the aircraft's ground speed and *distance made good.* For an airplane cruising at 250 knots airspeed, a 50-knot headwind will turn a 10-hour, 2,500-mile flight into a 12½-hour flight. There is a 20 percent loss of efficiency, and if the airplane is carrying 12 hours of fuel, it is not going to reach its destination. Contrariwise, a 50-knot headwind on the same flight for a jet cruising at 500 knots airspeed will result in only a 10 percent loss of efficiency and an increased flight time of one-half hour. The jet will require less fuel reserve to reach its destination.

In 1956, *Sovereign of the Skies* was the last commercial airplane to ditch on an ocean station[4] not because some special misfortune struck the aircraft but because commercial jets started taking over regular transoceanic service less than two years later. By the early 1970s, airlines did not attempt to cross the Atlantic or Pacific with piston-engine airplanes. Therefore, the need to have cutters on ocean stations to deal with ditchings disappeared.

VERSATILITY OF RESCUE AIRCRAFT. Use of a helicopter to hoist people off a sinking ship might appear to be easier than rescue by use of a ship, but that is not necessarily the case. A helicopter pilot must exercise incredible concentration to keep his aircraft in a hover above a ship whose masts are swinging wildly trying to bat the helicopter out of the sky. The helicopter hoist operator also must concentrate and guide the pilot, who cannot see where the hoist basket is going. A rescue swimmer who leaps from an aircraft into the open sea is taking obvious, great risks.

But a helicopter flies much faster than a ship sails, and so it can arrive on scene in minutes rather than hours. In frigid waters, that time is literally the difference between life and death.

Once technology led to helicopters capable of flying far offshore, the at-sea rescue of those who were wounded or ill or dying became the domain of the helicopter: It was faster than a ship, it put a fewer number of rescuers at risk, and it got the victims to full medical care much more quickly. For instance, in 1957 it took *Chincoteague* six days to travel carefully more than 1,000 miles through heavy seas to get the severely burned crewmen from *Finntrader* to a hospital; fifty years later, a Coast Guard helicopter made a flight of approximately 1,000 miles to pick up then bring the crewman of the *CSCL Genoa* to a hospital less than eight hours after he was first injured.

Helicopters of the twenty-first century have the ability to go much farther out to sea than could their predecessors because of increased speed and greatly improved navigation. An aircraft that flies at a higher rate of speed (e.g., a twenty-first century Coast Guard Jayhawk flies at 145 knots, whereas the Coast Guard's helicopter of the early 1960s, the Sikorsky Guardian, had a maximum speed of about 100 knots) will go farther on a load of fuel. And, as was proved by the *CSCL Genoa* case, it is possible for a helicopter to refuel at sea if a compatible ship is available. Finally, some, though not all, helicopters can conduct in-flight refueling with the result that there is no theoretical limit to how far offshore a helicopter can go.

Furthermore, the increased accuracy of navigation brought about by GPS means that a modern aircraft will fly directly to the scene of disaster without wasting time and fuel searching for a ship that is uncertain about its position. Time and fuel equate with distance offshore: Every pound of fuel saved from searching increases the range of the helicopter.

A helicopter as opposed to a flying boat can stay airborne while picking up survivors or injured people. The risk involved with a water landing is removed from the problem of completing the rescue. Therefore, hoisting rather than open-sea landing by flying boat or amphibious helicopter became routine for offshore recovery work.

Yet, not all offshore search-and-rescue work can be completed solely by helicopter. In October 1980, the *Hamilton*-class cutter *Boutwell* participated in the rescue of 319 passengers and 205 crewmen from the cruise ship *Prinsendam*, which had caught fire in the Gulf of Alaska approximately 130 miles south of Yakutat. Seven helicopters flew to the scene for the rescue operation, and most of the survivors were hoisted off the cruise ship's lifeboats by helicopter. To enable continuous air recovery of survivors as the helicopters filled with rescuees, the aircrew of a filled helicopter would fly the group of rescuees to the supertanker *Williamsburgh*, which happened to be transiting in the area, had two helipads, and was standing by. When one of the helicopters needed to leave the scene to refuel, its aircrew would take a load of rescuees and fly them to shore.[5] The air rescue operation might not have been a complete success had there not been ships on scene. This case proved that ships would still be required to deal with disaster at sea.

Mendota's boat crew returns with survivors from the burning Portuguese fishing schooner *Illhavense Segundo* on Ocean Station Delta, August 15, 1955.

* * *

UNITED STATES COAST GUARD cutters still go to sea in heavy weather. Cutters conducting fisheries patrols in the North Atlantic and the Bering Sea are pummeled by breaking seas accompanied by freezing temperatures. Cutters searching out smugglers in the Caribbean are threatened by ship-sinking hurricanes. Cutters at homeport may suddenly be called into the fury of the sea.

The legacy of four decades of at-sea stations is not the glorious cutters but rather the fortitude and seamanship of the cuttermen who make the cutters glorious.

ENDNOTES

1. Radiobeacons, which were developed during World War I, were in use along the U.S. coasts for the purpose of maritime navigation, but the effective range of the signal was only 200 to 300 miles offshore. The coastal radiobeacons were operated by the U.S. Lighthouse Service and became the responsibility of the U.S. Coast Guard when the Lighthouse Service was incorporated into the Coast Guard in 1939 (U.S. Coast Guard Headquarters, Office of Engineering, 1944, p. 2).

2. For the current network of buoys, see the National Data Buoy Center website: http://www.ndbc.noaa.gov/. One such buoy would be placed within fifty miles of the position of Ocean Station Hotel and was the final element in the elimination of the last ocean station. The buoy is called buoy Hotel.

3. At-sea weather reporting still does occur through the United States Voluntary Observing Ship (VOS) project operated by the National Data Buoy Center, National Weather Service, National Oceanic and Atmospheric Administration (NOAA) as part of the worldwide VOS Program. See the U.S. VOS website: http://www.vos.noaa.gov/.

4. In November 1957, nearly one year after the ditching of *Sovereign of the Skies*, a Pan Am clipper, *Romance of the Skies*, inexplicably crashed southwest of Ocean Station November. After an extensive search, U.S. Navy ships and the cutters *Bering Strait* and *Minnetonka* found no survivors.

5. Waters (1989, pp. 164–165).

APPENDIX I

Historical Document

The following is a transcription of the title page and the "Ocean Weather Observation Service" section of an undated (circa 1947), declassified topical history located in the archives of the U.S. Coast Guard Historian's Office, Washington, D.C. The transcribed material represents a contemporary account of the establishment of the Atlantic Weather Observation Service and of Stations 1 and 2.

~~CONFIDENTIAL~~

U. S. COAST GUARD

During the period July, 1939, to November 1, 1941, when the Coast Guard was transferred to the Navy, Headquarters was in the process of administering to a constantly expanding Service. On July 1, 1939, began the administration of a Coast Guard greatly enlarged by the consolidation with it of a long established civilian service. This presented Headquarters with the problem of integrating into its military establishment the personnel and material resources of a civilian service as old as the Coast Guard itself. While the task of perfecting this integration was in progress, the Coast Guard felt the full impact of important additional duties arising from the European War and later in preparing the Coast Guard for service with the Navy in World War II. It was a period of unparalleled activity in Service annals, and a general topical history of the more important phases of Headquarters administration throughout this time is given in the following accounts, viz:

Reorganization of Headquarters

Consolidation of the Lighthouse Service

Enforcement of neutrality

Control over the anchorage and movements of vessels, and security of ships and harbors

Greenland operations

Icebreaking

Ocean weather observation service

Coast Guard Reserve and Auxiliary

Transfer of the Coast Guard to the Navy

U.S. Maritime Service

Bureau of Marine Inspection and Navigation

Transfer of 10 cutters to the British Government

[. . .]

OCEAN WEATHER OBSERVATION SERVICE

The initial activities of the Coast Guard in carrying on a regular program of weather observations at sea, in connection with ocean weather forecasting in cooperation with the Weather Bureau, were carried on during the Ice Patrol season of 1939 in the North Atlantic. The Coast Guard cutters on International Ice Patrol duty from April 1, 1939, to July 1, 1939, made daily radiosonde observations of temperature, pressure, and humidity in the upper air over the Grand Banks region. The ocean stations at which these observations were made were part of the network over the North Atlantic area for experimental studies in the development of means for greater safety to life in trans-Atlantic aviation commerce. The data obtained, under the handicaps of no fixed location and taken as opportunity permitted in the course of the Ice Patrol duties, gave evidence of the importance of having data of this character and also the capacity and desire of the Coast Guard to work hand in hand with the Weather Bureau in furtherance of the development of a systematic and progressive weather service for ocean commerce.

Although weather forecasting was not basically a Coast Guard function, lying more within the jurisdiction of the Weather Bureau, the traditional Service spirit of cooperation with other Federal agencies in maritime affairs, the relationship of this program to safety at sea, led to Headquarters lending every encouragement to the Weather Bureau in broadening the field of ocean weather observation and to increasing its value to both shipping and air commerce. Close liaison was maintained between Headquarters and the Weather Bureau officials in an endeavor to work out a plan of joint action, with the Coast Guard furnishing vessels and personnel, and with the Weather Bureau providing certain technical weather observers and equipment. It was upon this joint community of interest that the foundation was laid for the present-day ocean weather observation service.

In the spring of 1939, about which time Pan-American Airways was undertaking the inaugural stage of regular trans-Atlantic air commerce, a series of conferences took place among the Civil Aeronautics Authority, the Weather Bureau, the Pan-American Airways, and the Coast Guard, regarding the "safety at sea" of overseas aircraft. One of the few points upon which there was mutual agreement was that it would be necessary, for safety of overseas aircraft, to station one or more ships in the Atlantic to take weather observations, that the Coast Guard was the proper agency to operate these ships; and that the Weather Bureau was the proper agency to furnish the technical personnel and equipment for observations.

In September, 1939, the President of the Pan-American Airways conveyed to the Commandant advice that the war situation had practically blanketed out weather reports from merchant vessels (French and British) in the Atlantic on which Pan-American had strongly depended in the past to insure safe service, and that unless something was done to insure an adequate and dependable weather reporting service over the ocean, it would be necessary for Pan-American Airways to suspend its trans-Atlantic service. He had also made representations to the White House and the Navy Department, and advised the Commandant that both strongly desired the continuance of the trans-Atlantic air commerce.

The President of Pan-American Airways proposed to the Commandant that the Coast Guard station one vessel on the northern route between Botwood and Foynes until the end of September and two vessels on the southern route between Bermuda and the Azores, and one vessel between the Azores and Lisbon throughout the year, for the purpose of collecting and reporting weather data and for safeguarding planes. While the Commandant viewed the proposal with full sympathy and wished to see the necessary measures taken, it was felt that Coast Guard craft were not entirely adapted to this service because of their small size and, furthermore, because of the additional burdens expected to be placed upon the Service in the enforcement of neutrality, it was not deemed practicable to assign any Coast Guard vessels to this duty except to a limited extent until other arrangements might be made.

While Headquarters could not see its way clear to assign Coast Guard cutters forthwith, the urgent need for such weather observation service and the recognition that the Coast Guard was the logical agency of the Government to perform the duty, led to the Commandant recommending to the Secretary of the Treasury that six former Shipping Board vessels be selected from those under the control of the Maritime Commission [and] be turned over to the Coast Guard for reconditioning and equipping for the service mentioned. It was contemplated that four would be required at sea and two as relief vessels. This recommendation was made after the Commandant had conferred with the Chief of the Weather Bureau, and it was agreed that the Service should be put in operation as soon as possible, with the Weather Bureau furnishing the necessary meteorological personnel and equipment. The Secretary of the Treasury viewed the proposal as meritorious, but felt the expense too great as to warrant the Treasury Department initiating action.

Subsequently on September 12, 1939, the Commandant had Rear Admiral Edward H. Smith, USCG, the Coast Guard's recognized authority on North Atlantic meteorological conditions, come to Washington from Boston, reviewing the problem in all its aspects in company with the Chief of the Weather Bureau and the Commandant. Aside from the representations of the Pan-American Airways, it was brought out that the Navy was considerably concerned over the scarcity of meteorological observations from the North Atlantic. It was thought that possibly, in the widespread desire to see this weather observation service established, the suggestion might be advanced that the squadron of destroyers-cutters then assigned to rescue operations in the vicinity of 55°W. be more widely deployed and undertake both the rescue and weather observation duties. To this proposal, if it should happen to be advanced in any quarters, the conference entertained strong and effective reasons in opposition and, on the contrary, felt that if a group of floating weather stations were established at strategic points in the North Atlantic Ocean, it could conceivably replace the destroyer-cutter squadron.

On October 12, 1939, Hon. H. A. Wallace, Secretary of Agriculture, under which Department the Weather Bureau operated, made formal written representations to both the Secretary of the Treasury and the Secretary of the Navy, in the interest of establishing this ocean weather observation service. Pertinent excerpts from this letter and one of similar purport under date of November 15, 1939, follows:

> "To some extent missing reports have been replaced by increasing the number of reports from vessels of American registry. However, there remain large areas in the Atlantic from which no reports are received. Under the circumstances the Weather Bureau [is] handicapped in giving warning of storms which affect our coastal regions. It is moreover difficult, if not

impossible, to provide an adequate aviation weather service for the maintenance of trans-ocean air traffic. One of the expected future developments in connection with the growth of trans-Atlantic flying is the assignment of a station vessel or vessels in critical areas to make observations and collect and forward reports for use in continental weather forecasting and for traffic control centers.* * *" "* *Weather reports from the Atlantic have taken on a new importance. They are needed not only for the general weather service of the country, but also because of their bearing on the safety of present trans-Atlantic aircraft operations.**" Unless the large gaps in the reporting system are corrected there exists grave danger during the winter season that trans-Atlantic planes will encounter disastrous conditions without warning. The Weather Bureau has made every effort to increase the number of vessel weather reports by arrangements with steamship lines, but it has found that such sources need to be supplemented under present international conditions by reports from ships stationed in the centers of areas from which reports are not now received. These vessels would transmit by radio not only their own observations of surface weather and upper air conditions, but would also be collecting centers to obtain reports from vessels in the region and forward the reports to Weather Bureau centers in the United States. The Weather Bureau would furnish the necessary meteorological staffs for these station vessels."

The Secretary of the Treasury replied that the Coast Guard was not in a position to use any of its vessels for this service. The Secretary of the Navy replied in similar tenor regarding naval vessels, but stated that it was felt this was a duty for the Coast Guard to perform, and suggested that suitable vessels might be made available by the Maritime Commission for transfer to the Coast Guard.

The need for this ocean weather observation service became so pressing that towards the end of the year (1939), the White House was brought into the picture through representations from both the Agriculture Department and the State Department. As a result the

Naval Aide to the President, at the direction of the President, had Admiral Stark, Chief of Naval Operations; Admiral Land, Chairman of the Maritime Commission; and Admiral Waesche meet in conference and make recommendations as to the availability of ships for the Atlantic Weather Patrol. The Naval Aide submitted these recommendations to the President on January 25, 1940, and the President indicated his approval of (a) "That Coast Guard cutters be withdrawn from the Grand Banks Patrol and established on the weather patrol duty. This would be a temporary measure until vessels are available which could be permanently assigned this duty", and (b) "Appropriations for Maritime Commission are presently being considered by Congress as part of the Independent Offices Bill. Should the funds be provided in this bill as asked by the Maritime Commission, Admiral Land will then be in a position to undertake the recommissioning of the four vessels for purpose of establishing a more permanent weather patrol, operated by the Coast Guard."

Thus was given the Executive authority for the establishment of the Atlantic Weather Observation Service. While throughout the preliminary negotiations stress was laid upon the need of these weather observations for the safety of trans-oceanic commerce, the Coast Guard also had kept in mind the importance of any vessel so assigned being capable of rendering assistance to any plane which might meet with mishap, and for maintaining a radio guard over the movements of the plane throughout its trans-ocean flight. The President, the Navy and Coast Guard also recognized the corollary value of vessels serving on these ocean outposts making observations of a national defense nature.

Headquarters, attaching great importance to this new undertaking, lost no time in drawing up plans and arranging for its inauguration. In order to make cutters available, it was necessary to discontinue the Grand Banks Patrol on January 27, 1940, to permit of the assignment of the cutters BIBB, CAMPBELL, DUANE, HAMILTON, INGHAM and SPENCER. These were all sister ships of the 327-foot class, the most modern and largest at the disposal of the Service. On February 1, 1940, Headquarters issued instructions to the Commanders of the Boston, New York and Norfolk Districts directing the commencement of the Atlantic Weather Observation Service as soon as practicable under the immediate direction of Headquarters. The original instructions setting up this service provided in part as follows:

"In cooperation with the U. S. Weather Bureau, two weather observation stations will be maintained, one in Latitude 35-38 N., longitude 53-21 W. (Station 1) and the other in Latitude 37-44 N., longitude 41-13 W. (Station 2). These stations are on the Great Circle course between St. Georges, Bermuda and Horta, Fayal Island, Azores, and are approximately one-third and two-thirds the distance between these points, respectively.

"These stations will be temporarily maintained by the BIBB, CAMPBELL, DUANE, HAMILTON, INGHAM and SPENCER, pending the refitting and availability of four Maritime Commission vessels, which will be permanently engaged in this duty and which will be manned by Coast Guard personnel.

"In order that the above-named cutters may be able to carry out these duties, it will be necessary to have installed on each, certain Weather Bureau equipment. There will also be assigned to each vessel, while engaged on the Weather Observation Stations, three representatives of the Weather Bureau to assist in this work.

"In view of the fact that the Weather Observation stations 1 and 2 are directly on the trans-Atlantic plane route being used at the present time (February, 1940), it is highly desirable that those stations be maintained as closely as circumstances will permit. While it is highly desirable that the judgment and initiative of the commanding officers be not interfered with, it is believed that leaving station to render assistance to surface craft should be reduced to the minimum and under circumstances which will not take the station vessel at any great distance from its station for any great length of time. Cases in which the need for assistance appears to be potential should be given careful consideration before any action involving leaving station is taken."

As a further explanation of the scope of work carried on by the ocean weather observation service, the following is quoted from instructions issued by Headquarters:

> "During the operation, all communications ship to shore will be handled with Coast Guard Radio, Washington. * * * The weather observations to be made include radio sonde, pilot balloon, and surface observations of the various meteorological elements. * * * Due to the seriousness of the international situation (1941) and radio silence imposed on the majority of vessels at sea in consequence thereof, communications with vessels will be very limited. However, vessels of the Atlantic Weather Observation Service may receive, upon request for further transmittal to Observer Washington, observations from vessels unable to communicate direct with shore radio stations. It is possible that the vessels while on station may receive requests for local weather and other advices. Such requests, if received by radio must be politely refused, as any transmission by radio in plain language of any meteorological information could be considered an unneutral act by a belligerent nation. Such requests may be answered if the reply may be transmitted direct by visual signalling methods or by megaphones.* * * While the primary purpose of the Atlantic Weather Observation Service is to obtain and transmit promptly to Observer, Washington, the meteorological data desired by the Weather Bureau, it is also important that trans-Atlantic aircraft be afforded a sure means of communications with the weather stations when they so desire. * * * In order to insure that the weather stations are aware that a trans-Atlantic

> aircraft is in flight and that a proper monitoring watch is maintained while the planes are in flight, all movements of all trans-Atlantic aircraft will be forwarded promptly to each weather station vessel. * * *"

With this project underway through utilization of regular Coast Guard cutters, Headquarters was anxious to obtain converted merchant vessels in order to release the cutters for patrol duties, the need for which was being intensified with the widening sphere of the war throughout the Atlantic area. The efficiency of these 327-foot cutters was well known to the Navy, and they could well fit into some of the undertakings which the Navy was planning or had underway.

In following out the original plan of using vessels under the control of the Maritime Commission, inspections were made in 1940 of some of the laid-up ships in the James River, for selection of several for conversion to ocean weather station duty. At this time submarine warfare was having its effect on available shipping of the belligerent nations, and the need for ships by belligerents was becoming more acute monthly. The result was that the Coast Guard had to give up the intention of acquiring merchant ships from those on the Atlantic Coast under the control of the Maritime Commission, and later turned to the Navy to acquire five freighters from the Great Lakes area from which area they could be better spared. The Coast Guard itself was without legal authority at this time to acquire and recondition merchant vessels in the interests of national defense, such as was possessed by the Navy Department, and it was for this reason that recourse was made to the good offices of the Navy in 1941.

By October, 1941, the Navy had acquired the merchant vessels AETNA (later the CGC MANASQUAN), the JOHN GEHM (later the CGC MENEMSHA), the J. FLOYD MASSEY (later the CGC MONOMOY) and the JOHN J. O'HAGAN and SAGINAW. These were all of the ocean-going freighter type, generally around 250 feet in length, and it was considered that, after repairs and refitting, could serve satisfactorily on the ocean weather stations. These were brought to the Atlantic coast for reconditioning, three being assigned to the Coast Guard and the Navy retaining two for Navy needs. Already in the progress of administration of the ocean weather observation service Headquarters had to arrange for the assignment of some of the cutters smaller than the 327-footers, as the Navy was pressing the Coast Guard for the release of several of the latter type for important naval missions and during the latter half of 1941 several of the 327-footers were placed under the operation of the Navy for duties other than weather observation.

With the Coast Guard feeling keenly the lack of sufficient cutters for the national defense tasks upon which it was called to perform, efforts were made in 1941, to acquire, under the provisions of the Coast Guard Reserve and Auxiliary Act, some suitable large-size private yachts which could be brought into use on the weather patrol in addition to the ships being required by the Navy, and thus relieve the cutters for active patrol duties. The yacht NOURMAHAL was offered to the Service by its owner, and was inducted into the Coast Guard Reserve fleet on August 21st, and subsequently dispatched to the Coast Guard Yard, Curtis Bay, Maryland, for reconditioning for ocean weather service.

By August, 1941 the need for additional ocean weather stations in the North Atlantic became pressing, and was being urged for the area generally northeast of Newfoundland. The Weather Bureau had termed this area a "blind spot" where several disturbances could develop without the forecaster being able to anticipate them unless he had synoptic reports. The proposal for these two additional weather stations had been favorably considered by the Secretary of the Navy and the Defense Meteorological Committee, predicated upon providing protection to increasing aircraft operations over the Atlantic, including patrol, ferrying and commercial air transport. The expansion was also being urged by the British Government in representations through the State Department for the United States Government to take over several of the British stations on the trans-Atlantic route between Newfoundland and the United Kingdom.

The Coast Guard was disposed to undertake the operation of these additional stations provided the ships themselves were made available by the Navy or Maritime Commission. The availability of merchant ships, adapted for this service, had now become a problem due to the general ship shortage, and by the time war broke out in

December, 1941, no suitable vessels had been made available to the Coast Guard. Later in January, 1942, the yacht SEA CLOUD was acquired by the Navy and assigned to the Ocean Weather Patrol.

The establishment of the two ocean weather stations by the Coast Guard in 1940 in the mid-Atlantic was the basis of the expansion which followed in this field during the progress of the war, when ocean weather stations were established in the North Atlantic, South Atlantic and Pacific.

[*Nota bene:* In official documents related to the World War II–era stations, terms such as Atlantic Weather Observation Patrol, Atlantic Weather Observation Service, Atlantic Weather Patrol, North Atlantic Weather Patrol, Ocean Weather Patrol, Ocean Weather Station, Weather Observation Patrol, Weather Observation Station, Weather Station, and Weather Patrol are used indiscriminately or interchangeably.]

APPENDIX II

Historical Memorandum

The following is a transcription of the memorandum signed by President Franklin D. Roosevelt on January 25, 1940, to establish the Atlantic Weather Observation Service. (Source: U.S. Coast Guard Historian's Office, Washington, D.C.)

COPY

THE WHITE HOUSE

WASHINGTON

January 25, 1940

MEMORANDUM FOR

THE PRESIDENT

Subject: Ships for Atlantic Weather Observation

Pursuant to the President's memorandum to me, Admiral Stark, Admiral Land and Admiral Waesche have conferred on the matter of availability of ships for the Atlantic Weather patrol. As a result of this conference, the following is recommended:

(a) That Coast Guard cutters be withdrawn from the Grand Banks Patrol and established on the weather patrol duty. This would be a temporary measure until vessels are available which could be permanently assigned to this duty.

(b) Appropriations for Maritime Commission are presently being considered by Congress as part of the Independent Offices Bill. Should the funds be provided in this bill as asked by the Maritime Commission, Admiral Land will then be in a position to undertake the recommissioning of the four vessels for purpose of establishing a more permanent weather patrol, operated by the Coast Guard.

(c) If funds under (b) are reduced so that money will not be available for recommissioning, the question of additional legislation for this purpose will be taken up.

Does the President desire that (a), above, be carried out and that the work of recommissioning be undertaken if funds become available as outlined in (b), above?

Respectfully,

(Signed) D. J. Callaghan

Yes

FDR

APPENDIX III

Station Cutters 1940–1977

Class Descriptions	Unique
Active: 230 tons, 125 feet LOA, 1,200 horsepower	Aivik: 590 tons (displ.), 128 feet LOA, 525 boat horsepower
Algonquin: 1,000 tons, 165 feet LOA, 1,500 horsepower	Asterion: 6,600 tons (displ.), 382 feet LOA, 1,200 horsepower
Cactus: 1,000 tons, 180 feet LOA, 1,200 horsepower	Big Horn: 15,400 tons (full load), 425 feet LOA, 3,300 horsepower
Casco: 2,500 tons, 311 feet LOA, 6,000 horsepower	Manhasset ex-Wilton†
Edsall: 1,490 tons, 306 feet LOA, 6,000 horsepower	Muskeget ex-Cornish†
Hamilton: 3,000 tons, 378 feet LOA, 7,000 (diesel) horsepower; 36,000 (turbine) horsepower	Northland: 2,150 tons, 216 feet LOA, 1,000 horsepower
	Nourmahal: 3,200 tons (displ.), 263 feet LOA, 3,200 horsepower
Lakes type*: 2,580 tons (LDT), 261 feet LOA, 1,200 horsepower	Sea Cloud: 3,000 tons (displ.), 316 feet LOA, 2,700 horsepower
Owasco: 1,980 tons, 255 feet LOA, 4,000 horsepower	
Secretary: 2,350 tons, 327 feet LOA, 6,200 horsepower	
Tacoma: 2,200 tons, 303 feet LOA, 5,500 horsepower	
Tampa: 1,950 tons, 240 feet LOA, 2,700 horsepower	

*The Lakes-type freighters were not the Lake-class cutters. The Coast Guard referred to these freighters as Lakes type as they were built in 1918 to the same design and were originally named after lakes.

†The specifications for Manhasset and for Muskeget could not be determined.

Name(s)*	Class	Coast Guard Hull Number(s)	Date of Coast Guard Commissioning	Date of Coast Guard Decommissioning	Disposition
Abilene	Tacoma	PF-58	13-Mar-1946	21-Aug-1946	Sold to The Netherlands
Absecon	Casco	WAVP-374 WHEC-374	1-May-1949	9-May-1972	Transferred to Republic of South Vietnam
Active	Active	WSC-125	30-Nov-1926	2-Apr-1962	Sold
Aivik	Unique	WYP-164	24-Jun-1942	21-Jun-1944	Returned to owner
Alexander Hamilton; ex-Hamilton; ex-Alexander Hamilton	Secretary	WPG-34	4-Mar-1937	30-Jan-1942	Sunk by torpedo
Algonquin	Algonquin	WPG-75	20-Oct-1934	18-Apr-1947	Sold for scrap
Androscoggin	Owasco	WPG-68 WHEC-68	26-Sep-1946	27-Feb-1973	Sold for scrap
Asterion	Unique	WAK-123	12-Jan-1944	20-Jul-1944	Sold for scrap
Bangor	Tacoma	PF-16	15-Apr-1946	2-Aug-1946	Sold to Mexico
Barataria	Casco	WAVP-381 WHEC-381	10-Jan-1949	29-Aug-1969	Sold for scrap
Bering Strait	Casco	WAVP-382 WHEC-382	14-Dec-1948	1-Jan-1971	Transferred to Republic of South Vietnam

Name(s)*	Class	Coast Guard Hull Number(s)	Date of Coast Guard Commissioning	Date of Coast Guard Decommissioning	Disposition
Bibb; ex-George M. Bibb	Secretary	WPG-31 WAGC-31 WHEC-31	10-Mar-1937	30-Sep-1985	Artificial reef, Florida Keys
Big Horn	Unique	WAO-124	1-Jan-1944	1-Feb-1945	Returned to U.S. Navy for service
Boutwell	Hamilton	WHEC-719	24-Jun-1968	In Service as of 2009	In Service as of 2009
Brownsville	Tacoma	PF-102	15-Apr-1946	2-Aug-1946	Scrapped
Campbell; ex-George W. Campbell	Secretary	WPG-32 WAGC-32 WHEC-32	16-Jun-1936	1-Apr-1982	Sunk as a target
Casco	Casco	WAVP-370 WHEC-370	19-Apr-1949	21-Mar-1969	Sunk as a target
Castle Rock	Casco	WAVP-383 WHEC-383	16-Sep-1948	21-Dec-1971	Transferred to Republic of South Vietnam
Chambers	Edsall	WDE-491	11-Jun-1952	30-Jul-1954	Returned to U.S. Navy for service; subsequently sold for scrap
Chase	Hamilton	WHEC-718	11-Mar-1968	In Service as of 2009	In Service as of 2009
Chautauqua	Owasco	WPG-41 WHEC-41	4-Aug-1945	1-Aug-1973	Sold for scrap

Chincoteague	Casco	WAVP-375 WHEC-375	7-Mar-1949	9-May-1972	Transferred to Republic of South Vietnam
Comanche	Algonquin	WPG-76	1-Dec-1934	29-Jul-1947	Sold to the Virginia Pilots Association
Conifer	Cactus	WAGL-301 WLB-301	1-Jul-1943	2-Jun-2000	Sold for private use
Cook Inlet	Casco	WAVP-384 WHEC-384	15-Jan-1949	21-Dec-1971	Transferred to Republic of South Vietnam
Coos Bay	Casco	WAVP-376 WHEC-376	4-May-1949	1-Sep-1966	Sunk as a target
Corpus Christi	Tacoma	PF-44	15-Apr-1946	5-Jul-1946	Sold
Covington	Tacoma	PF-56	16-Mar-1946	17-Sep-1946	Sold to Ecuador
Dallas	Hamilton	WHEC-716	26-Oct-1967	In Service as of 2009	In Service as of 2009
Dexter; ex-USS Biscayne	Casco	WAGC-385 WAVP-385 WHEC-385	20-Sep-1946	18-Jan-1968	Sunk as a target
Duane; ex-William J. Duane	Secretary	WPG-33 WAGC-6 WHEC-33	1-Aug-1936	1-Aug-1985	Artificial reef, Florida Keys
Durant	Edsall	WDE-489	9-May-1952	10-Apr-1954	Returned to U.S. Navy for service; subsequently sold for scrap
Escanaba (I)	Algonquin	WPG-77	23-Nov-1932	13-Jun-1943	Exploded and sank

Name(s)*	Class	Coast Guard Hull Number(s)	Date of Coast Guard Commissioning	Date of Coast Guard Decommissioning	Disposition
Escanaba (II); ex-Ostego	Owasco	WPG-64 WHEC-64	20-Mar-1946	28-Jun-1974	Uncertain; believed scrapped
Evergreen	Cactus	WAGL-295 WLB-295 WAGO-295 WMEC-295	30-Apr-1943	13-Jun-1990	Sunk as a target
Falgout	Edsall	WDE-424	24-Aug-1951	18-Dec-1953	Returned to U.S. Navy for service; subsequently sunk as target
Finch	Edsall	WDE-428	24-Aug-1951	23-Apr-1954	Returned to U.S. Navy for service, subsequently sold for scrap
Forster	Edsall	WDE-434	20-Jun-1951	25-May-1954	Returned to Navy for service; subsequently transferred to the Republic of South Vietnam
Forsyth	Tacoma	PF-102	15-Mar-1946	2-Aug-1946	Returned to U.S. Navy; subsequently sold to The Netherlands

Frederick Lee	Active	WSC-139	4-Apr-1927	15-Dec-1964	Sold
Gallatin	Hamilton	WHEC-721	9-May-1969	In Service as of 2009	In Service as of 2009
Gladwyne	Tacoma	PF-62	15-Apr-1946	31-Aug-1946	Sold to Mexico
Gresham; ex-USS Willoughby	Casco	WAVP-387 WHEC-387 WAGW-387	1-Dec-1947	25-Apr-1973	Sold for scrap
Groton	Tacoma	PF-29	13-Mar-1946	27-Aug-1946	Sold to Colombia
Haida	Tampa	WPG-45	26-Oct-1921	13-Feb-1947	Sold then subsequently scrapped
Half Moon	Casco	WAVP-378 WHEC-378	14-Sep-1948	15-Jul-1969	Sold for scrap
Hamilton	Hamilton	WHEC-715	18-Mar-1967	In Service as of 2009	In Service as of 2009
Humboldt	Casco	WAVP-372 WHEC-372	24-Jan-1949	30-Sep-1969	Sold for scrap
Hutchinson	Tacoma	PF-45	15-Apr-1946	23-Sep-1946	Sold to Mexico
Ingham; ex-Samuel D. Ingham	Secretary	WPG-35 WAGC-35 WHEC-35	12-Sep-1936	27-May-1988	Museum ship in Key West, Fla.
Iroquois	Owasco	WPG-43 WHEC-43	9-Feb-1946	13-Jan-1965	Sold for scrap
Jarvis	Hamilton	WHEC-725	17-Jan-1972	In Service as of 2009	In Service as of 2009
Klamath	Owasco	WPG-66 WHEC-66	19-Jun-1946	1-May-1973	Sold for scrap

Name(s)*	Class	Coast Guard Hull Number(s)	Date of Coast Guard Commissioning	Date of Coast Guard Decommissioning	Disposition
Koiner	Edsall	WDE-431	20-Jun-1951	14-May-1954	Returned to U.S. Navy for service; subsequently sold for scrap
Lansing	Edsall	WDE-488	11-Jun-1952	29-Mar-1954	Returned to U.S. Navy for service; subsequently sold for scrap
Laurel	Cactus	WAGL-291 WLB-291	24-Nov-1942	11-Jul-1986	Sold for private use
Lowe	Edsall	WDE-425	20-Jul-1951	5-Apr-1954	Returned to U.S. Navy for service; subsequently sold for scrap
Mackinac	Casco	WAVP-371 WHEC-371	1-Apr-1949	28-Dec-1967	Sunk as target
Manasquan	Lakes-type	WAG-273	2-Apr-1942	22-Feb-1945	Sold
Manhasset	Unique	WAG-276	8-Aug-1942	15-Oct-1945	Sold
Manitowoc	Tacoma	PF-61	14-Mar-1946	3-Sep-1946	Sold to France
Matagorda	Casco	WAVP-373 WHEC-373	8-Jun-1949	15-Oct-1967	Sunk as target

McCulloch; ex-USS Wachapreague	Casco	WAVP-386 WHEC-386	25-Nov-1946	9-May-1972	Transferred to Republic of South Vietnam
Mellon	Hamilton	WHEC-717	9-Jan-1968	In Service as of 2009	In Service as of 2009
Mendota	Owasco	WPG-69 WHEC-69	2-Jun-1945	1-Nov-1973	Sold for scrap
Menemsha	Lakes-type	WAG-274	22-Oct-1943	24-Sep-1945	Sold
Midgett	Hamilton	WHEC-726	30-Mar-1972	In Service as of 2009	In Service as of 2009
Milledgeville	Tacoma	PF-94	7-May-1946	16-Aug-1946	Scrapped
Minnetonka; ex-Sunapee	Owasco	WPG-67 WHEC-67	11-Jul-1946	31-May-1974	Scrapped
Moberly; ex-Scranton	Tacoma	PF-63	15-Apr-1946	12-Aug-1946	Sold for scrap
Mohawk	Algonquin	WPG-78	19-Jan-1935	8-Jan-1948	Museum ship in Key West, Fla.
Mojave	Tampa	WPG-47	12-Dec-1921	3-Jul-1947	Sold
Monomoy	Lakes-type	WAG-275	22-Oct-1943	12-Oct-1945	Scrapped
Morgenthau	Hamilton	WHEC-722	24-Jul-1969	In Service as of 2009	In Service as of 2009
Munro	Hamilton	WHEC-724	15-Apr-1972	In Service as of 2009	In Service as of 2009
Muskeget	Unique	WAG-48	30-Jun-1942	9-Sep-1942	Sunk by torpedo
Muskegon	Tacoma	PF-24	15-Mar-1946	27-Aug-1946	Sold to France

Name(s)*	Class	Coast Guard Hull Number(s)	Date of Coast Guard Commissioning	Date of Coast Guard Decommissioning	Disposition
Newell	Edsall	WDE-422	20-Jul-1951	14-Apr-1954	Returned to U.S. Navy for service; subsequently sold for scrap
Northland	Unique (gunboat)	WPG-49	7-May-1927	27-Mar-1946	Sold
Nourmahal	Unique (yacht)	WPG-72 WPG-122	21-Aug-1941	30-May-1946	Sold for scrap
Orange	Tacoma	PF-43	15-Apr-1946	28-Oct-1946	Sold for scrap
Owasco	Owasco	WPG-39 WHEC-39	18-May-1945	27-Jun-1973	Sold for scrap
Pontchartrain; ex-Okeechobee	Owasco	WPG-70 WHEC-70	28-Jul-1945	19-Oct-1973	Uncertain; believed scrapped
Ramsden	Edsall	WDE-482	28-Mar-1952	10-Apr-1954	Returned to U.S. Navy for service, subsequently sunk as target
Richey	Edsall	WDE-485	28-Mar-1952	14-May-1954	Sunk as a target
Rockaway	Casco	WAVP-377 WHEC-377 WAGO-377 WOLE-377	24-Dec-1948	26-Sep-1966	Sold for scrap
Rush	Hamilton	WHEC-723	25-Sep-1969	In Service as of 2009	In Service as of 2009

Sea Cloud	Unique (yacht)	WPG-284	4-Apr-1942	9-Apr-1943	Privately owned charter ship as of 2009
Sebago; ex-Wachusett	Owasco	WPG-42 WHEC-42	20-Sep-1945	29-Feb-1972	Transferred to MARAD
Sheboygan	Tacoma	PF-57	14-Mar-1946	17-Sep-1946	Sold to Belgium
Sherman	Hamilton	WHEC-720	3-Sep-1968	In Service as of 2009	In Service as of 2009
Sorrel	Cactus	WAGL-296 WLB-296	15-Apr-1943	20-Jun-1996	Sold for private use
Spencer; ex-John C. Spencer	Secretary	WPG-36 WAGC-36 WHEC-36	1-Mar-1937	15-Dec-1980	Sold for scrap
Tahoma	Algonquin	WPG-80 WAGE-80	22-Oct-1934	1-Jul-1953	Sold for scrap
Taney; ex-Robert B. Taney	Secretary	WPG-37 WAGC-37 WHEC-37	24-Oct-1936	7-Dec-1988	Museum ship in Baltimore, Md.
Unimak	Casco	WAVP-379 WHEC-379	3-Jan-1949	29-Apr-1988	Artificial reef, Virginia coast
Vance	Edsall	WDE-487	9-May-1952	3-Apr-1954	Returned to U.S. Navy for service; subsequently sunk as target
Wachusett; ex-Huron	Owasco	WPG-44 WHEC-44	23-Mar-1946	30-Aug-1973	Sold for scrap

Name(s)*	Class	Coast Guard Hull Number(s)	Date of Coast Guard Commissioning	Date of Coast Guard Decommissioning	Disposition
Winnebago	Owasco	WPG-40 WHEC-40	21-Jun-1945	27-Feb-1973	Sold for scrap
Winona	Owasco	WPG-65 WHEC-65	19-Apr-1946	31-May-1974	Sold for scrap
Woonsocket	Tacoma	PF-32	16-Mar-1946	18-Sep-1946	Transferred to Peru
Yakutat	Casco	WAVP-380 WHEC-380	23-Nov-1948	1-Jan-1971	Transferred to Republic of South Vietnam

*This table lists those ships commissioned as Coast Guard cutters that patrolled the weather observation stations, plane guard stations, or ocean stations. Thus, of the Tacoma-class frigates (Navy-commissioned ships manned by Coast Guard crews during World War II), only those that were later commissioned as cutters are listed.

BIBLIOGRAPHY

Various sources provided general information on U.S. Coast Guard history, the merchant marine, ship operations, and the ocean station program; they are listed in the "General" section of the Bibliography. Specific source documents for each chapter are then listed by chapter.

GENERAL

Books and Periodicals

Baker, Elijah, *Introduction to Steel Shipbuilding*, McGraw-Hill, New York, NY, 1953.

Blackman, Raymond V.B. (editor), *Jane's Fighting Ships 1970–71*, McGraw-Hill, New York, NY, 1970.

Canney, Donald L., *U.S. Coast Guard and Revenue Cutters, 1790–1935*, Naval Institute Press, Annapolis, MD, 1995.

Capron, Walter C., Captain, USCG (Ret.), *U.S. Coast Guard*, Franklin Watts, New York, NY, 1965.

Couhat, Jean Labayle, *Combat Fleets of the World 1982/3*, Naval Institute Press, Annapolis, MD, 1982.

Crenshaw, R.S., Jr., Captain, USN (Ret.), *Naval Shiphandling*, 4th ed., Naval Institute Press, Annapolis, MD, 1975.

Drummond, John D., *Gap of Danger: The Story of the Weather Ships*, W.H. Allen, London, UK, 1963.

Dunn, Laurence, *The World's Tankers*, Adlard Coles Limited, London, UK, 1956.

Floherty, John J., *Search and Rescue at Sea*, J.B. Lippincott, Philadelphia, PA, 1953.

Harkins, Philip, *Coast Guard Ahoy*, Harcourt, Brace, New York, NY, 1943.

Hayler, William B. (editor-in-chief) and Keever, John M. (associate editor), *American Merchant Seaman's Manual*, 17th ed., Cornell Maritime Press, Centreville, MD, 2003.

Johnson, Robert Erwin, *Guardians of the Sea: History of the United States Coast Guard 1915 to the Present*, Naval Institute Press, Annapolis, MD, 1987.

King, Irving, H., *The Coast Guard Expands, 1865–1915: New Roles, New Frontiers*, Naval Institute Press, Annapolis, MD, 1996.

LaDage, John H., *Modern Ships: Elements of Their Design, Construction, and Operations*, 2nd ed., Cornell Maritime Press, Centreville, MD, 1965.

LaDage, John, and Van Gemert, Lee (George, William E., editor), *Stability and Trim for the Ship's Officer*, 3rd ed., Cornell Maritime Press, Centreville, MD, 1983.

Larzalere, Alex, *The Coast Guard at War, Vietnam 1965–1975*, Naval Institute Press, Annapolis, MD, 1997.

Marton, Greg S., *Tanker Operations: A Handbook for the Ship's Officer*, Cornell Maritime Press, Centreville, MD, 1978.

Meurn, Robert J., *Survival Guide for the Mariner*, Cornell Maritime Press, Centreville, MD, 1993.

Milwee, William I., Jr., *Modern Marine Salvage*, Cornell Maritime Press, Centreville, MD, 1996.

Moore, John E., Captain, RN (editor), *Jane's Fighting Ships 1975–76*, Franklin Watts, New York, NY, 1975.

Murphy, L.J., CDR, Commanding Officer, CGC *Duane*, *USCGC Duane*, privately published book, Portland, ME, ca. August 1985.

Nalty, Bernard C., and Strobridge, Truman R., "The Story of the Ocean Stations," *Alumni Bulletin*, Vol. XXVI, No. 2., U.S. Coast Guard Academy, New London, CT, 1974.

Nalty, Bernard C., Noble, Dennis L., and Strobridge, Truman R. (editors), *Wrecks, Rescues and Investigations*; Scholarly Resources, Inc., Wilmington, DE, 1978.

Noble, Dennis L., *That Others Might Live: The U.S. Life-Saving Service, 1878–1915*, Naval Institute Press, Annapolis, MD, 1994.

Noble, Dennis L., and Wilson, Elwyn E., "End of an Era," *Mariners Weather Log*, National Oceanic and Atmospheric Administration, Washington, DC, ca.1978.

Noel, John V., Jr., *Knight's Modern Seamanship*, 17th ed., Van Nostrand Reinhold, New York, NY, 1984.

O'Connor, Neil F., "OSVs: End of an Era," *Proceedings*, U.S. Naval Institute, Annapolis, MD, November 1973.

Pearcy, Arthur, *U.S. Coast Guard Aircraft Since 1916*, Naval Institute Press, Annapolis, MD, 1991.

Powell, Hickman, *What the Citizen Should Know About the Coast Guard*, W.W. Norton, New York, NY, 1941.

Rachlis, Eugene, *The U.S. Coast Guard*, Random House, New York, NY, 1961.

Scheina, Robert L., *U.S. Coast Guard Cutters and Craft of World War II*, Naval Institute Press, Annapolis, MD, 1982.

Scheina, Robert L., *U.S. Coast Guard Cutters and Craft 1946–1990*, Naval Institute Press, Annapolis, MD, 1990.

Strobridge, Truman R., and Noble, Dennis L., *Alaska and the U.S. Revenue Cutter Service 1867-1915*, Naval Institute Press, Annapolis, MD, 1999.

U.S. Naval Oceanographic Office, *Distances Between Ports*, U.S. Government Printing Office, Washington, DC, 1965.

Waters, John M., Jr., Captain, USCG, *Rescue at Sea*, D. Van Nostrand, Princeton, NJ, 1966; Naval Institute Press, Annapolis, MD, 1989 (2nd ed.).

Waters, John M., Jr., Captain, USCG (Ret.), *Bloody Winter*, revised ed., Naval Institute Press, Annapolis, MD, 1984.

Willoughby, Malcom F., *Rum War at Sea*, U.S. Government Printing Office, Washington, DC, 1964.

Willoughby, Malcom F., *The U.S. Coast Guard in World War II*, Naval Institute Press, Annapolis, MD, 1957; Naval Institute Press, Annapolis, MD, 1989 (rev. printing).

Government Documents

Capelotti, Pete, PAC, USCG; "Oceanography in the Coast Guard," undated pamphlet, U.S. Coast Guard, Washington, DC.

Craik, J.D., Deputy Commander, Eastern Area, U.S. Coast Guard, "SAR Study on Coast Guard assistance rendered during the Severe Atlantic storm of 14–20 Nov 1962," New York, NY, 28 January 1963.

Dinsmore, R.P., Captain, USCG (Ret.), "Ocean Weather Ships 1940–1980." U.S. Coast Guard Historian's Office, Washington, DC. Available at: http://www.uscg.mil/history/webcutters/rpdinsmore_oceanstations.asp. Accessed 1 June 2009.

Hammond, R.E., Chief, Office of Operations, U.S. Coast Guard, "Plan for Closing of VICTOR, ECHO and WHEC Decommissioning," official Coast Guard memo, Washington, DC, 15 November 1971.

International Civil Aviation Organization, "ICAO Ocean Stations at Work," *ICAO Bulletin*, Vol. XV, No. 9, International Civil Aviation Organization, Montreal, Canada, 1960.

Moritz, G.H., Chief, Shipbuilding and Maintenance Branch, U.S. Coast Guard, "WHEC Recommissioning," official Coast Guard memo, Washington, DC, 24 September 1974.

Naval History & Heritage Command. *Dictionary of American Naval Fighting Ships*. Naval History & Heritage Command, Washington Navy Yard, DC. Available at: http://www.history.navy.mil/danfs.

Naval Sea Systems Command (Department of the Navy), *Naval Vessel Register, The Official Inventory of Naval Ships and Service Craft*, Washington Navy Yard, Washington, DC. Available at: http://www.nvr.navy.mil.

Reed, James A., Assistant Secretary of the Treasury, *Coast Guard Roles and Missions; Volume III, Ocean Stations*, Washington, DC, September 1962.

Stika, J.E., Commander Western Area, U.S. Coast Guard, *Standard Operating Procedure for Pacific Ocean Stations*, San Francisco, CA, 1950.

U.S. Coast Guard Commandant (Chief of Staff), "Decommissioning of High Endurance Cutters during FY 1973," Official Coast Guard message, Washington, DC, 16 January 1973.

U.S. Coast Guard Headquarters Public Affairs Division, "History of USCGC *Dexter*," Washington, DC, undated.

U.S. Coast Guard Headquarters Public Information Division, *The Coast Guard at War; Lost Cutters, VIII*, Washington, DC, 1 June 1949.

U.S. Coast Guard Headquarters Public Information Division, *The Coast Guard at War; Transports and Escorts, V*, Washington, DC, 1 June 1949.

U.S. Coast Guard Headquarters Public Information Division, *The Coast Guard at War; Weather Patrol, VII*, Washington, DC, 1 June 1949.

U.S. Coast Guard Headquarters Public Information Division, "USCGC *Chincoteague*," Washington, DC, undated (ca. October 1957).

U.S. Coast Guard Headquarters Public Information Division, *Weathermen of the Sea: Ocean Station Vessels*, Washington, DC, 1961.

U.S. Coast Guard Historian's Office, Washington, DC, disaster files for: *Fort Mercer, Helga Bolten, Keo, Lampsis, Monte Palomares, Pamir, Pendleton, Smith Voyager, Sovereign of the Skies, Texaco Oklahoma, Viking Princess.*

U.S. Coast Guard Historian's Office, Washington, DC, ship's files for cutters: *Absecon, Active, Androscoggin, Asterion, Barataria, Bering Strait, Bibb, Big Horn, Boutwell, Campbell, Casco, Castle Rock, Chambers, Chase, Chautauqua, Chincoteague, Cook Inlet, Coos Bay, Dallas, Dexter, Duane, Durant, Escanaba, Evergreen, Falgout, Forster, Forsyth, Frederick Lee, Gallatin, Gresham, Half Moon, Hamilton, Humboldt, Ingham, Iroquois, Jarvis, Klamath, Koiner, Lansing, Laurel, Lowe, Mackinac, Manasquan, Manhasset, Matagorda, McCulloch, Mellon, Mendota, Menemsha, Midgett, Minnetonka, Monomoy, Morgenthau, Munro, Muskeget, Newell, Northland, Nourmahal, Owasco, Pontchartrain, Ramsden, Richey, Rockaway, Rush, Sea Cloud, Sebago, Sherman, Sorrel, Spencer, Taney, Unimak, Vance, Wachusett, Winnebago, Winona, Yakutat.*

U.S. Coast Guard Historian's Office, Washington, DC, subject files for the Ocean Station Program.

U.S. Coast Guard Public Information Division, *Weathermen of the Sea; Ocean Station Vessels*, Washington, DC, 1961.

U.S. Coast Guard Special Issue Study, *The Ocean Station Vessel Program After 1973* (three volumes), Washington, DC, June 1971.

CHAPTER 1

Books and Periodicals

Parrott, Daniel S., *Tall Ships Down: The Last Voyages of the Pamir, Albatross, Marques, Pride of Baltimore, and Maria Asumpta*, International Marine/McGraw-Hill, Camden, ME, and New York, NY, 2003.

Paterson, Tony, "German Ship Which Sank Killing 80 Was 'Rust Bucket' That Should Not Have Sailed," *Independent* (UK), 22 June 2007. Available at: http://www.independent.co.uk/news/world/europe/german-ship-which-sank-killing-80-was-rust-bucket-that-should-not-have-sailed-454159.html. Accessed 1 June 2009.

Schäuffelen, Otmar, *Chapman Great Sailing Ships of the World*, Hearst Books, New York, NY, 2005.

Villiers, Alan, *The Way of a Ship*, Charles Scribners Sons, New York, NY, 1953.

Villiers, Alan, *The War with Cape Horn*, Charles Scribners Sons, New York, NY, 1971.

Walters, Sydney D., *Pamir – The Story of a Sailing Ship*, A.H. and W. Reed, Wellington, New Zealand, 1949.

Willoughby, Malcom F., The *U.S. Coast Guard in World War II*, Naval Institute Press, Annapolis, MD, 1989 (revised printing of the 1957 edition).

Government Documents

Imlay, M.H., Captain, USCG (Air Sea Rescue Officer, Eastern Sea Frontier), "New Beacon Calls and Position Reporting Grid System for Atlantic Weather Patrol Ships," Letter ESF93/A6-3/H2(S), Ser. 7982, New York, NY, 18 September 1945.

Moore, Paul L. (and staff), "The Hurricane Season of 1957," *Monthly Weather Review*, National Weather Bureau, Washington, DC, December 1957.

Stika, J.E., Commander Western Area, U.S. Coast Guard, *Standard Operating Procedure for Pacific Ocean Stations*, San Francisco, CA, 1950.

U.S. Coast Guard Cutter *Absecon*, Official Logbook, September 1957a.

U.S. Coast Guard Cutter *Absecon*, "Report of Assistance," Norfolk, VA, 8 October 1957b.

U.S. Coast Guard Headquarters Public Information Division, *The Coast Guard at War; Weather Patrol, VII*, Washington, DC, 1 June 1949.

West Germany Federal Minister of Trade, *Decisions of the Federal Supreme Admiralty Court and of the Admiralty Courts of the German Confederation. Volume 5, Parts 1-5, Admiralty Court of Lubeck. The sinking of the Sailing Schoolship Pamir*, Bonn and Hamburg, 1958.

CHAPTER 2

Books and Periodicals

Hefley, Gifford, *The Development of Loran-C Navigation and Timing* (National Bureau of Standards Monograph 129), U.S. Department of Commerce, Washington, DC, 1972.

Kauffman, Sanford B. (George Hopkins, ed.), *Pan Am Pioneer: A Manager's Memoir from Seaplane Clippers to Jumbo Jets*, Texas Tech University Press, Lubbock, TX, 1995.

Lawrence, Harry, *Aviation and the Role of Government*, Kendall Hunt, Dubuque, IA, 2004.

MacTaggart-Cowan, P.D., "Transatlantic Aviation and Meteorology," *Journal of the Royal Astronomical Society of Canada*, Vol. XXXII, No. 5, May–June 1938, pp. 217–231.

Milner, Samuel, "Establishing the Bolero Ferry Route," *Military Affairs*, Vol. 11, No. 4, Winter 1947, pp. 213–222.

Morison, Samuel Eliot, *The Battle of the Atlantic, 1939–1943* (History of United States Naval Operations in World War II, Volume I), Little, Brown and Company, Boston, MA, 1957. (Note: The 1957 printing is the revised and corrected edition of the original 1947 printing of Volume I of the series. The companion volume is Volume X, *The Battle of the Atlantic, 1943–1945*.)

Pierce, J.A., McKenzie, A.A., and Woodward, R.H., editors, *Loran: Long Range Navigation* (Massachusetts Institute of Technology Radiation Laboratory Series, Volume 4; Office of Scientific Research and Development, National Defense Research Committee). McGraw-Hill, New York, NY, 1948.

Reynolds, Clark, G., "Voyage of the NC-3," *Naval History*, Vol. 1, No. 1, April 1987.

Rohwer, Jürgen, *Axis Submarine Successes, 1939–1945*, Naval Institute Press, Annapolis, MD, 1983.

Rohwer, Jürgen, *Chronology of the War at Sea, 1939–1945: The Naval History of World War Two*, Naval Institute Press, Annapolis, MD, 2005.

Wiktor, Christian L., *Multilateral Treaty Calendar/Répertoire des traités multilatéraux, 1648–1995*, Martinus Nijhoff/Kluwer Law, The Hague, The Netherlands, 1998.

Wilbur, T., Commander, USN, "The First Flight Across the Atlantic, May 1919," originally published in *Naval Aviation News*, republished by the

National Air and Space Museum, Smithsonian Institution, Washington, DC, 1969. Available from the Naval History & Heritage Command at: http://www.history.navy.mil/branches/nc-4mono.htm. Accessed 1 June 2009.

Williams, Frances Leigh, *Matthew Fontaine Maury, Scientist of the Sea*, Rutgers University Press, New Brunswick, NJ, 1963.

Willoughby, Malcom F., The *U.S. Coast Guard in World War II*, Naval Institute Press, Annapolis, MD, 1989 (revised printing of the 1957 edition).

Government Documents

Ahearn, J.L., Captain, USCG, Chief, Military Morale Section, U.S. Coast Guard, "Albert Adam Kelsch," official Coast Guard letter to Miss Nan Kelsch, Washington, DC, 1 November 1942.

Bernier, G.N., Commanding Officer, USS *Sea Cloud*, "Rescue of Survivors, Portuguese fishing schooner '*Maria Da Gloria*,'" official letter; Boston, MA, 17 August 1942.

Boussard, L.C., Public Information Officer, "Third ICAO Conference on North Atlantic Ocean Stations, Brighton, July 1953," ICAO News Release, International Civil Aviation Organization, Montreal, Canada, 28 July 1953.

Chalker, L.T., Rear Admiral, Assistant Commandant, U.S. Coast Guard, "AG-48, ex-*Cornish*; Commissioning as *Muskeget*," letter to Prospective Commanding Officer, *Muskeget*, Washington, DC, 24 June 1942.

Chalker, L.T., Rear Admiral, Assistant Commandant, U.S. Coast Guard, "*Monomoy*, relief from Weather Observation Patrol duty," letter to Commanding Officer, *Monomoy*, Washington, DC, 14 July 1943.

Chalker, L.T., Rear Admiral, Assistant Commandant, U.S. Coast Guard, "Atlantic Weather Patrol," letter to Commanding Officer, *Manhasset*, Washington, DC, 18 March 1944a.

Chalker, L.T., Rear Admiral, Assistant Commandant, U.S. Coast Guard, "CGC *Manasquan* (ex-*Aetna*); recommendation for disposition of," letter to Commander in Chief U.S. Fleet; Washington, DC, 19 November 1944b.

Dreyer, William A.C., Lieutenant Commander, USNR, Commandant Third Naval District, "Turnover of USS *Muskeget*," New York, NY, 30 June 1942.

Foley, E.H., Acting Secretary of the Treasury, "Operation of Ocean Weather Stations," letter to Vice Admiral Land, President, Air Transport Association of America, Washington, DC, 30 July 1948.

Furer, Julius Augustus, Rear Admiral, USN (Ret.), *Administration of the Navy Department in World War II*, U.S. Government Printing Office, Washington, DC, 1959.

Heiner, J.N., Captain, USCG, Chief Naval Engineering Division, Coast Guard Headquarters, "Disposition of *Evelyn*, formerly *CG-Asterion*," official Coast Guard letter, Washington, DC, 31 May 1945.

Horne, F.J., Vice Chief of Naval Operations; "CGC *Manasquan* (ex-*Aetna*) – Recommendation for disposition of," official Navy letter, Washington, DC, 9 December 1944.

IOC/WMO (Intergovernmental Oceanographic Commission/World Meteorological Organization), *Oceanographic Aspects of the First GARP Global Experiment (FGGE)*. IOC/INF-351. UNESCO, Paris, 1978.

Lister, Ernest A., "The North Atlantic Ocean Stations Agreement," *Department of State Bulletin*, Vol. XXX, No. 778, Washington, DC, 24 May 1954. (Mr. Lister was the chairman of the U.S. delegation to the Fourth Conference on North Atlantic Ocean Stations.)

Olsen, John L., Commanding Officer, Coast Guard Receiving Station, "Decommissioning (CGC *Menemsha*)," official Coast Guard letter, Norfolk, VA, 26 September 1945.

Owens, C.T., Captain, USN (Ret.), District Operations Officer, First Naval District, "USS *Asterion* transferred to Coast Guard," official Navy Speedletter, Boston, MA, 12 January 1944.

Provisional International Civil Aviation Organization, "Final Act and International Agreement on North Atlantic Weather Observation," London, United Kingdom, 25 September 1946. [Published in Hearings before the Subcommittee on Coast Guard of the Committee on Merchant Marine and Fisheries, House of Representatives, Eightieth Congress, Washington, DC, 8 July 1947.]

Reinburg, LeRoy, Captain, USCG, Commandant, Coast Guard Yard, "*Monomoy*; orders, movement, assignment to duty, and permanent station," official Coast Guard letter, Curtis Bay, MD, 15 October 1943.

Shelley, Tully, Chief of Naval Operations (By direction), "Yacht *Sea Cloud*, Assignment of," Washington, DC, 2 February 1942.

Toba, Y., Hanawa, K., and Kurasawa, Y., "Examination of Time Series Data at OWS TANGO." In: *Time Series of Ocean Measurements*, Volume I

(Intergovernmental Oceanographic Commission Technical Series No. 24), UNESCO, Paris, 1983.

United Nations, *Treaty Series*, No. 900, "Exchange of Notes Constituting an Interim Agreement Between the United States of America and Canada Relating to the Establishment of a Network of Seven Weather Stations in the Pacific Ocean. Washington, 8 and 22 June 1950." Vol. 70, p. 115, 1950 (70 UNTS 115).

United Nations, *Treaty Series*, No. 900, "Exchange of Notes Constituting an Interim Agreement Between the United States of America and Canada Relating to the Establishment of a Network of Seven Weather Stations in the Pacific Ocean. Washington, 8 and 22 June 1950." Vol. 87, p. 390, 1951 (87 UNTS 390).

United Nations, *Treaty Series*, No. 1405, "North Atlantic Ocean Weather Stations Agreement. Signed at London, on 12 May 1949." Vol. 101, p. 91, 1951 (101 UNTS 91).

United Nations, *Treaty Series*, No. 1405, "North Atlantic Ocean Weather Stations Agreement. Signed at London, on 12 May 1949." (Protocol Extending the Period of the Above-Mentioned Agreement. Done in Montreal, on 28 May 1952.) Vol. 150, p. 380, 1952 (150 UNTS 380).

United Nations, *Treaty Series*, No. 2922, "Final Act of the Fourth International Civil Aviation Organization Conference on North Atlantic Ocean Stations. Done at Paris, on 25 February 1954; and Agreement on North Atlantic Ocean Stations. Done at Paris, on 25 February 1954." Vol. 215, p. 249, p. 268, 1955 (215 UNTS 249, 268).

United States Air Force and United States Navy, "Flight Information Publication, Enroute Supplement, Pacific and Far East," 7651st Aeronautical Chart and Information Squadron, U.S. Air Force, San Francisco, CA, 16 October 1961.

U.S. Coast Guard Commandant, "Operating Plan Change Order 56-70, redeployment of CGC *Gresham* (WHEC-387)," Washington, DC, 29 January 1970.

U.S. Coast Guard Commandant, "Operating Plan Change Order 53-72, Ocean Station Vessel Program reductions," Washington, DC, 14 January 1972.

U.S. Coast Guard Cutter *Bibb*, Official Logbook, February 1940.

U.S. Coast Guard Cutter *Duane*, Official Logbook, February 1940.

U.S. Coast Guard Cutter *Duane*, Official Logbook, June 1941.

U.S. Coast Guard Cutter *Laurel*, Vessel Movement Record, Boston, MA, and Portland, ME, 1943–1945.

U.S. Coast Guard Cutter *Sorrel*, Vessel Movement Record, Boston, MA, 1943–1945.

U.S. Coast Guard Headquarters, Office of Engineering, *Loran, Volume I, Early Electronic History and the Bridging of the North Atlantic and North Pacific*, Washington, DC, December 1944 [U.S. Coast Guard Headquarters Public Information Division, *The Coast Guard at War, Loran, IV, Volume I*, Washington, DC, 1 May 1946].

U.S. Coast Guard Headquarters Public Information Division, *The Coast Guard at War; Transports and Escorts, V*, Washington, DC, 1 June 1949a.

U.S. Coast Guard Headquarters Public Information Division, *The Coast Guard at War; Weather Patrol, VII*, Washington, DC, 1 June 1949b.

U.S. Coast Guard Headquarters Public Information Division, *The Coast Guard at War; Lost Cutters, VIII*, Washington, DC, 1 June 1949c.

U.S. Coast Guard Historian's Office, Washington, DC, ship's files for cutters *Monomoy*, *Muskeget*, and *Escanaba*.

U.S. Coast Guard Official Dispatch 041759, "USS *Sea Cloud* Placed Out of Commission," Boston, MA, 4 November, 1944.

U.S. Coast Guard Public Information Division, Office of Public Relations, "History of USS *Sea Cloud* (IX 99)," Washington, DC, 31 March 1949.

Waesche, R.R., Commandant, U.S. Coast Guard, "U.S.S. *Menemsha* (ex-S.S. *John Gehm*); commissioning; orders," letter to Prospective Commanding Officer, U.S.S. *Menemsha*, Washington, DC, 12 December 1941a.

Waesche, R.R., Commandant, U.S. Coast Guard, "U.S.S. *Monomoy;* commissioning; orders," letter to Prospective Commanding Officer, U.S.S. *Monomoy*, Washington, DC, 12 December 1941b.

Waesche, R.R., Commandant, U.S. Coast Guard, "*Manasquan* (ex-S.S. *Aetna*); commissioning; orders," letter to Prospective Commanding Officer, *Manasquan*, Washington, DC, 12 January 1942a.

Waesche, R.R., Commandant, U.S. Coast Guard, "U.S.S. *Monomoy;* orders; amendment," letter to Commanding Officer, U.S.S. *Monomoy*, Washington, DC, 24 January 1942b.

Waesche, R.R., Commandant, U.S. Coast Guard, "Yacht *Sea Cloud*," letter to the Chief of Naval Operations, Washington, DC, 24 January 1942c.

CHAPTER 3

Government Documents

Bailey, C.W., Commanding Officer CGC *Coos Bay*; "The *Ambassador* Story," official Coast Guard narrative, Portland, ME, 24 February 1964.

Coffin, E.A., Commanding Officer, CGC *Sebago*, "Steadying sail, comments on," official Coast Guard letter, Staten Island, NY, 28 December 1948.

Coffin, E.A., Commanding Officer, CGC *Sebago*; "Excessive rolling; installation gyro stabilizers, recommendation for," official Coast Guard letter, Staten Island, NY, 12 January 1949.

Gerde, C.S., Chief, Naval Engineering Division, Coast Guard Headquarters, "*McCulloch* (WAVP-386); Storm Damage," official Coast Guard memorandum, Washington, DC, 3 March 1959.

Richmond, A.C., Commandant, U.S. Coast Guard, "*McCulloch* Storm Damage, January 1959," official Coast Guard letter, Washington, DC, 15 April 1959.

U.S. Coast Guard Cutter *Coos Bay*, Official Logbook, February 1964.

U.S. Coast Guard Cutter *McCulloch*, Official Logbook, January 1959.

U.S. Coast Guard Cutter *Owasco*, Official Logbook, January 1962.

U.S. Coast Guard Historian's Office, Coast Guard Medal Award Citation for David P. Birchcrest. U.S. Coast Guard Historian's Office, Washington, DC.

CHAPTER 4

Books and Periodicals

Rohwer, Jürgen, *Chronology of the War at Sea, 1939–1945: The Naval History of World War Two*, Naval Institute Press, Annapolis, MD, 2005.

Taylor, Philips V., "Rescue on the North Atlantic," *The Quarterdeck Log*, Vol. 21, No. 2, Coast Guard Combat Veterans Association, Westfield Center, OH, Summer 2007.

Government Documents

Bostock, W.H., and 47 Passengers of the *Bermuda Sky Queen*; "Testimony of the Survivors of the Flying Boat, '*Bermuda Sky Queen*,'" undated letter (ca. October 1947) accompanying the Official Logbook of the Coast Guard Cutter *Bibb*, National Archives, Washington, DC.

Civil Aeronautics Board, Accident Report - NC18612, Accident Investigation Report, "American International Airways, Inc – Bermuda Sky Queen, North Atlantic Ocean – October 14, 1947" (SA-152, File No. 1-0088). Washington, DC, 15 December 1948.

Cronk, Paul B., Commanding Officer, CGC *Bibb*, "Letter to Members of the U.S. Merchant Marine," official Coast Guard letter, At Sea, 19 October 1947.

Foley, E.H., Jr, Acting Secretary of the Treasury, "Citation to accompany the award of the Treasury Department Gold Lifesaving Medal to Boatswain's Mate First Class Marion K. Reynolds," Washington, DC, 2 March 1948.

Riley, LT F.X., Board of Awards Secretary, "Information Concerning the Rescue of 69 Survivors of the Bermuda Sky Queen by Personnel of the USGCG *Bibb*," Washington, DC, undated (ca. November 1947).

Snyder, John W., Secretary of the Treasury, "Citation to accompany the award of the Treasury Department Gold Lifesaving Medal to LTJG C.S. Hall," Washington, DC, undated (ca. 1948).

Snyder, John W., Secretary of the Treasury, "Citation to accompany the award of the Treasury Department Silver Lifesaving Medal to Boatswain's Mate Ralph Keller," Washington, DC, undated (ca. 1948).

Snyder, John W., Secretary of the Treasury, "Citation to accompany the award of the Treasury Department Silver Lifesaving Medal to Gunner's Mate John Johnson," Washington, DC, undated (ca. 1948).

Snyder, John W., Secretary of the Treasury; "Citation to accompany the award of the Treasury Department Silver Lifesaving Medal to LTJG B.S. Brown," Washington, DC undated (ca. 1948).

Snyder, John W., Secretary of the Treasury, "Citation to accompany the award of the Treasury Department Silver Lifesaving Medal to Machinist Harry LaFever," Washington, DC, undated (ca. 1948).

Snyder, John W., Secretary of the Treasury, "Citation to accompany the award of the Treasury Department Silver Lifesaving Medal to Machinist Lindel Hall," Washington, DC, undated (ca. 1948).

Snyder, John W., Secretary of the Treasury, "Citation to accompany the award of the Treasury Department Silver Lifesaving Medal to Machinist Philips Van Campen Taylor," Washington, DC, undated (ca. 1948).

Sprow, N.W., Military Morale Division, Coast Guard Headquarters, "Letter to Mr. Leon M. Siler concerning BM1 Marion K. Reynolds," Washington, DC, 4 October 1948.

U.S. Coast Guard Commander, Fifth District, "Background Data *Chincoteague* Mid-ocean Transfer of Injured Men from Freighter *Finntrader* 11 January 1957," Norfolk, VA, undated (ca. January 1957).

U.S. Coast Guard Cutter *Bibb*, Official Logbook, October 1947 and September 1948.

U.S. Coast Guard Cutter *Chincoteague*, Official Logbook, January 1957.

U.S. Coast Guard Cutter *McCulloch*, Official Logbook, September 1947.

CHAPTER 5

Government Documents

Malen, J.W., Commanding Officer, CGC *Barataria*, "USCGC *Barataria*; commissioning of," official Coast Guard letter, Mare Island, CA, 8 January 1949.

Richmond, A.C., Commandant, U.S. Coast Guard, "*McCulloch* Storm Damage, January 1959," official Coast Guard letter, Washington, DC, 15 April 1959.

U.S. Coast Guard Commander, Fifth District, "Brief Resume of the Cutter *Chincoteague* Involving the SS *Helga Bolten*," Norfolk, VA, undated (ca. November 1956).

U.S. Coast Guard Cutter *Chincoteague*, "Report of Assistance," Norfolk, VA, 26 November 1956.

U.S. Coast Guard Cutter *Chincoteague*, Official Logbook, October 1956 and January 1957.

U.S. Coast Guard Cutter *Half Moon*, Official Logbook, October 1949.

U.S. Coast Guard Cutter *McCulloch*, Official Logbook, September 1947; February 1952; January 1959.

U.S. Coast Guard Headquarters Public Affairs Division, "History of USCGC *Dexter*," Washington, DC, undated.

U.S. Coast Guard Public Affairs Office, "Press Packet for Presentation of a Plaque to the Coast Guard Cutter *Chincoteague*," Norfolk, VA, undated (ca. October 1957).

Wayne, Albert F., Commanding Officer, CGC *McCulloch*, "*McCulloch* (WAVP-386) Storm Damage, January 1959, Report of," official Coast Guard letter, Boston, MA, 19 January 1959.

CHAPTER 6

Books and Periodicals

Cadigan, J.J., "Decision at Sea: Rescue of Crewmen from the *Smith Voyager*," *Readers' Digest* XC, January 1967, pp. 50–54.

Elphick, Peter, *Liberty: The Ships that Won the War*, Naval Institute Press, Annapolis, MD, 2006.

Hayes, Evelyn. "Dr. Constance Tipper: Testing her mettle in a materials world," *Advanced Materials & Processes/ASM News*, Vol. 153, No. 6, 1998.

LaDage, John H., *Modern Ships – Elements of Their Design, Construction, and Operations*, 2nd ed., Cornell Maritime Press, Centreville, MD, 1965.

Lane, Frederic C. (with the collaboration of Blanche D. Coll, Gerald J. Fischer, and David B. Tyler), *Ships for Victory: A History of Shipbuilding Under the U.S. Maritime Commission in World War II*, The Johns Hopkins Press, Baltimore, MD, 1951.

Sawyer, L.A., and Mitchell, W.H., *The Liberty Ships*, 2nd ed., Lloyds of London Press, New York, NY, 1985.

Government Documents

Fowler, Henry H., Secretary of the Treasury, "Citation to accompany the award of the Silver Life-Saving Medal to Donald James Covert," Washington, DC, 22 October 1965a.

Fowler, Henry H., Secretary of the Treasury, "Citation to accompany the award of the Silver Life-Saving Medal to James A. Wright," Washington, DC, 22 October 1965b.

Fowler, Henry H., Secretary of the Treasury, "Citation to accompany the award of the Silver Life-Saving Medal to Eckart Linke," Washington, DC, 15 February 1967.

U.S. Coast Guard Commander, Eastern Area, Press Release *Monte Palomares*, New York, NY, 10 January 1966a.

U.S. Coast Guard Commander, Eastern Area, Press Release *Monte Palomares*, New York, NY, 11 January 1966b.

U.S. Coast Guard Commander, Eastern Area, Press Release *Monte Palomares*, New York, NY, 12 January 1966c.

U.S. Coast Guard Cutter *Androscoggin*, Official Logbook, January 1966a.

U.S. Coast Guard Cutter *Androscoggin*, "Situation Reports SS *Lampsis*" (numbers one through six, nine, and eleven), official Coast Guard messages, At Sea, January 1966b.

U.S. Coast Guard Cutter *Barataria*, Official Logbook, January 1966.

U.S. Coast Guard Cutter *Escanaba*, Official Logbook, January 1966a.

U.S. Coast Guard Cutter *Escanaba*, "Situation Reports M/V *Monte Palomares*" (numbers one through nine), official Coast Guard messages, At Sea, January 1966b.

U.S. Coast Guard Cutter *Rockaway*; Official Logbook, December 1964.

U.S. Coast Guard Cutter *Spencer*, Official Logbook, December 1964.

U.S. Coast Guard Historian's Office, Coast Guard Medal Award Citations for Roger C. Boggs and John J. Cadigan. U.S. Coast Guard Historian's Office, Washington, DC

U.S. Coast Guard Marine Board of Investigation, *SS Smith Voyager, sinking of in North Atlantic Ocean, 28-30N, 50-48W, on 27 December 1964 while under tow following abandonment on 20 December 1964 at 29-45N, 48-50W, with the loss of life of four crewmembers during rescue operations*, Washington, DC, 11 June 1965.

CHAPTER 7

Books and Periodicals

Chapline, W.E., Jr., "Towing Seaplanes," *U.S. Coast Guard Academy Alumni Association Bulletin*, Vol. X, No. 4, U.S. Coast Guard Academy, New London, CT, July 1948.

Frump, Robert, *Two Tankers Down: The Greatest Small-Boat Rescue in U.S. Coast Guard History*, The Lyons Press, Guilford, CT, 2008.

Webber, Bernard C., *Chatham, The Lifeboatmen*, Lower Cape Publishing, Orleans, MA, 1985.

Webster, W. Russell, Captain, USCG, "The *Pendleton* Rescue," *Proceedings*, U.S. Naval Institute, Annapolis, MD, December 2001.

Government Documents

Mathieu, D.J., Commanding Officer, CGC *Half Moon*, CGC *Half Moon* Instruction No. 1050.2, "Liberty and Recall Procedures while on Search and Rescue Standby Duty in Bermuda," Staten Island, NY, 7 November 1964.

Paetzel, Frederick C., Sworn Statement concerning the rescue of the crew of S.S. *Fort Mercer* by CGC *Yakutat*, 25 February 1952. (Document located in the National Archives, Record Group 26, Lifesaving Medal Folder for ENS William R. Kiely.)

Renshaw, L.W.A., Commanding Officer, CGC *Rockaway*, "Narrative Navy P5M #127711," official Coast Guard report, Staten Island, NY, 1958.

Snyder, John W., Secretary of the Treasury, "Citation to accompany the award of the Treasury Department Gold Lifesaving Medal to ENS W.R. Kiely," Washington, DC, 7 May 1952.

Turner, Edward E., Sworn Statement concerning the rescue of the crew of S.S. *Fort Mercer* by CGC *Yakutat*, New York, NY, 27 February 1952.

U.S. Coast Guard Board of Awards, "Recommendation for Award of Gold and Silver Lifesaving Medals for ENS Kiely, ENS Carmichael, EN2 Black, SA Mason, and SN Terwilliger," Washington, DC, 31 March 1952.

U.S. Coast Guard Commandant, First Endorsement to MVI memorandum of 25 September 1952 (Fort Mercer - a-1 Bd), "Marine Board of Investigation; structural failure of tanker FORT MERCER off Cape Cod on 18 February 1952, with loss of life," 17 October 1952; Approved 20 October 1952, Washington, DC, 1952a.

U.S. Coast Guard Commandant, First Endorsement to MVI memorandum of 25 September 1952 (Pendleton - a-1 Bd), "Marine Board of Investigation; structural failure of tanker PENDLETON off Cape Cod on 18 February 1952, with loss of life," 17 October 1952; Approved 20 October 1952, Washington, DC, 1952b.

U.S. Coast Guard Cutter *Acushnet*, Official Logbook, February 1952.

U.S. Coast Guard Cutter *Androscoggin*, Official Logbook, May 1952.

U.S. Coast Guard Cutter *Casco*, Official Logbook, February 1956.

U.S. Coast Guard Cutter *Eastwind*, Official Logbook, February 1952.

U.S. Coast Guard Cutter *McCulloch*, Official Logbook, February 1952.

U.S. Coast Guard Cutter *Rockaway*, Official Logbook, September 1958.

U.S. Coast Guard Cutter *Yakutat*, Official Logbook, February 1952.

CHAPTER 8

Books and Periodicals

Price, Scott T., Assistant Coast Guard Historian, "The Forgotten Service in the Forgotten War: The U.S. Coast Guard's Role in the Korean Conflict," *The Quarterdeck Log*, Vol. 15, No. 4, Coast Guard Combat Veterans Association, Westfield Center, OH, Winter 2001.

Government Documents

Earle, William K., Commanding Officer, CGC *Pontchartrain*, "Ditching of Pan American Clipper 10943 at Ocean Station November on 16 October 1956," official Coast Guard report, Long Beach, CA, 18 October 1956.

French, Robert Lewis, Aviation Electronicsman First Class, U.S. Navy, "Twin Crash at Sea," official statement, United States Pacific Fleet, Patrol Squadron Twenty-Two, San Francisco, CA, undated (ca. 1953).

Smith, Willard J., Commandant, U.S. Coast Guard, *Aircraft Emergency Procedures Over Water, CG-306*, U.S. Department of Transportation, Washington, DC, 1968.

U.S. Coast Guard Commander, Fourteenth District, Press Release 58-3-22, Honolulu, HI, September 1958.

U.S. Coast Guard Cutter *Chambers*, Vessel Movement Record, New Bedford, MA, 1952–1954.

U.S. Coast Guard Cutter *Coos Bay*, Official Logbook, February 1953 and January 1955.

U.S. Coast Guard Cutter *Durant*, Vessel Movement Record, Honolulu, HI, 1952–1954.

U.S. Coast Guard Cutter *Falgout*, Vessel Movement Record, Tacoma, WA, 1951–1953.

U.S. Coast Guard Cutter *Finch*, Vessel Movement Record, Alameda, CA, 1952–1954.

U.S. Coast Guard Cutter *Forster*, Vessel Movement Record, Honolulu, HI, 1952–1954.

U.S. Coast Guard Cutter *Gresham*, Official Logbook, December 1958.

U.S. Coast Guard Cutter *Lansing*, Vessel Movement Record, Seattle, WA, 1952–1954.

U.S. Coast Guard Cutter *Lowe*, Vessel Movement Record, Seattle, WA, 1951–1954.

U.S. Coast Guard Cutter *Koiner*, Vessel Movement Record, Seattle, WA, 1951–1954.

U.S. Coast Guard Cutter *Newell*, Vessel Movement Record, Honolulu, HI, 1951–1954.

U.S. Coast Guard Cutter *Pontchartrain*, Official Logbook, October 1956.

U.S. Coast Guard Cutter *Ramsden*, Vessel Movement Record, Seattle, WA, 1952–1954.

U.S. Coast Guard Cutter *Richey*, Vessel Movement Record, Honolulu, HI, 1952–1954.

U.S. Coast Guard Cutter *Sebago*, Official Logbook, April 1949.

U.S. Coast Guard Cutter *Vance*, Vessel Movement Record, Seattle, WA, 1952–1954.

U.S. Coast Guard Cutter *Winona*, Official Logbook, September 1952; October 1952; November 1952; December 1952.

U.S. Coast Guard Historian's Office, "Coast Guard Lifesaving Medal Rescue," Joseph R. Bridge, Winfield J. Hammond, Robert F. Hewitt, Joseph M. Miller, Jr., Tracy W. Miller, Gerald W. Stuart, Carle R. Tornell, and John Vukic, Washington, DC.

U.S. Coast Guard Historian's Office, Subject file for Coast Guard PBM aircraft, Washington, DC.

U.S. Coast Guard Historian's Office, Subject files for the Korean War, Washington, DC.

U.S. Department of Commerce, *Ocean Station Instructions for Meteorological Personnel (Supplementary)*, U.S. Department of Commerce, Washington, DC, April 1950.

CHAPTER 9

Books and Periodicals

Amerson, A. Binion, Jr., *The Natural History of French Frigate Shoals, Northwestern Hawaiian Islands*, Atoll Research Bulletin No. 150 (Paper No. 79, Pacific Ocean Biological Survey Program), Smithsonian Institution, Washington, DC. Available at: http://www.sil.si.edu/digitalcollections/atollresearchbulletin/issues/00150x.pdf. Accessed 1 June 2009.

Walter, Robert, Simons, Gerald, Gossner, Simone D., and Horan, Anne, *Hawaii: The American Wilderness*, Time-Life Books, New York, NY, 1973.

Uhlig, Frank, Jr., *How Navies Fight: The U.S. Navy and Its Allies*, Naval Institute Press, Annapolis, MD, 1994.

Government Documents

Commander, U.S. Naval Base Guantanamo, "Distress M/V *Viking Princess*," official Navy message, Guantanamo, Cuba, April 1966.

Kolkhorst, LCDR B.E., "Transfer of Command, USCGC *Chautauqua* (WPG-41); report of," official Coast Guard letter, Honolulu, HI, 29 July 1954.

Tulich, Eugene N., "The United States Coast Guard in South East Asia During the Vietnam Conflict," monograph, U.S. Coast Guard Historian's Office, Washington, DC, undated (ca. 1998).

U.S. Coast Guard Commander, Fourteenth District, "Operation Order No. 58-54," Honolulu, HI, 25 August 1954.

U.S. Coast Guard Commander, Seventh District, "Situation Reports M/V Viking Princess Afire and Abandoning Ship" (numbers one through five), official Coast Guard messages, Miami, FL, April 1966.

U.S. Coast Guard Cutter *Barataria*, Official Logbook, March 1968.

U.S. Coast Guard Cutter *Cook Inlet*, Official Logbook, April 1966.

U.S. Coast Guard Cutter *Iroquois*, Official Logbook, June 1954 and July 1954.

U.S. Coast Guard Cutter *Matagorda*, Official Logbook, January 1965; February 1965; March 1965; April 1965.

U.S. Coast Guard Headquarters, "Weekly Report of Activities and Developments," Numbers 26 through 34, Washington, DC, July–September 1954.

U.S. Coast Guard Historian's Office, "Coast Guard Medal Award Citations," for Jon Thorvald Norstog, Washington, DC.

U.S. Coast Guard Historian's Office, "Coast Guard Medal Award Citations," for Robert S. Barnes, Albert C. Buechler, Albert C. Davies, Ronald J. Dinsmore, John E. Johnson, and William O'Keefe, Washington, DC.

CHAPTER 10

Government Documents

Adams, W.F., Commander, Twelfth Coast Guard District (by direction), "*Barataria's* Decommissioning, Key Schedule," official Coast Guard letter, San Francisco, CA, 17 July 1969.

Adams, W.F., Commander, Twelfth Coast Guard District (by direction), "CGC *Gresham* Decommissioning; Key Schedule," official Coast Guard letter, San Francisco, CA, 9 September 1969.

Allen, E.C., Commander, Fifth Coast Guard District, "SAR Case Study, TEXACO TANKER OKLAHOMA [sic]," Portsmouth, VA, 20 July 1971.

Editor and Staff, "Case Summary – *Texaco Oklahoma*," *On Scene – The National Maritime SAR Review*, U.S. Coast Guard Headquarters, Washington, DC, September 1972.

McMullen, I.H., Chief of Staff, Third Coast Guard District, "SAR Case Study of the loss of M/T *Keo*/ELAQ (LI) and Yacht *Whisper*," New York, NY, 22 January 1970.

National Transportation Safety Board, *Structural Failure and Sinking of the Texaco-Oklahoma Off Cape Hatteras on 27 March 1971, With the Loss of 31 Lives*, Washington, DC, 24 May 1972.

Schuman, S.M., Commanding Officer, CGC *Campbell*, "M/T *KEO*, SAR incident, evaluation of; report concerning," official Coast Guard letter, Portland, ME, 15 December 1969.

Shuhy, J.L., *Oceanographic Observations: North Atlantic Ocean Station Bravo, 14 November 1967–16 November 1969* (Oceanographic Report No. CG 373-63), United States Coast Guard Oceanographic Unit, Washington, DC, April 1974.

Siler, Owen W., Commandant, U.S. Coast Guard, Clearance Sheet, "WHEC Disposition," to accompany U.S. Coast Guard (G-CPA/84 7100 ELT) memo dated 11 January 1978, Washington, DC.

Skillings, J.P., Commanding Officer, USCGC *Gresham*, "News release; CGC *Gresham* decommissioning," official Coast Guard letter, Norfolk, VA, 10 April 1973.

U.S. Coast Guard Commandant, "Operating Plan Change Order 56-70, redeployment of CGC *Gresham* (WHEC-387)," Washington, DC, 29 January 1970.

U.S. Coast Guard Commandant, Operating Facility Change Order 86-73, Decommissioning of CGC *Gresham* (WAGW-387), Washington, DC, 15 February 1973.

U.S. Coast Guard Commandant (Chief of Staff), "Decommissioning of CGC *Gresham*," official Coast Guard message, Washington, DC, 4 November 1969.

U.S. Coast Guard Commandant (Chief of Staff), "Decommissioning of High Endurance Cutters during FY 1973," official Coast Guard message, Washington, DC, 16 January 1973.

U.S. Coast Guard Commandant (Personnel), "Gresham Personnel Planning," official Coast Guard message, Washington, DC, 8 January 1970.

U.S. Coast Guard Commander, Fifth District, "Distress *Texaco Oklahoma*," official Coast Guard message, Portsmouth, VA, 29 March 1971a.

U.S. Coast Guard Commander, Fifth District, "Situation Report Two *Texaco Oklahoma*," official Coast Guard message, Portsmouth, VA, 29 March 1971b.

U.S. Coast Guard Commander, Fifth District, "Situation Report Three *Texaco Oklahoma*," official Coast Guard message, Portsmouth, VA, 29 March 1971c.

U.S. Coast Guard Commander, Fifth District, "Situation Report Five *Texaco Oklahoma*," official Coast Guard message, Portsmouth, VA, 30 March 1971d.

U.S. Coast Guard Commander, Fourteenth District, "Public Affairs Release 36-74; Cutter *Mellon* to Close Ocean Station November," Honolulu, HI, 19 May 1974.

U.S. Coast Guard Commander, Twelfth District, "(*Gresham*) Decom Plans," official Coast Guard message, San Francisco, CA, 30 September 1969a.

U.S. Coast Guard Commander, Twelfth District, "*Gresham* Decommissioning Schedules," official Coast Guard message, San Francisco, CA, 3 October 1969b.

U.S. Coast Guard Commander, Twelfth District, "Crypto/Comtac support for *Gresham* Redeployment," official Coast Guard message, San Francisco, CA, 5 January 1970.

U.S. Coast Guard Commander, Twelfth District, "Public Affairs Release 3-72; Pacific Ocean Station 'Victor' Will be Discontinued," San Francisco, CA, 12 January 1972.

U.S. Coast Guard Cutter *Campbell*, Official Logbook, November 1969.

U.S. Coast Guard Cutter *Campbell*, "Oscar Sierra Delta Buried at Sea," official Coast Guard message, At Sea, 30 June 1973.

U.S. Coast Guard Cutter *Gresham*, Official Logbook, February 1970.

U.S. Coast Guard Cutter *Yakutat*, Official Logbook, November 1969.

U.S. Coast Guard Headquarters, "Public Affairs Release 19-73; Discontinuing Three mid-Atlantic and one mid-Pacific weather stations," Washington, DC, 14 March 1973.

U.S. Coast Guard Headquarters, "Public Affairs Release 51-74; Coast Guard Scheduled to End 34 Year Old Ocean Station Program," Washington, DC, 28 June 1974.

U.S. Coast Guard Headquarters Public Affairs Division, "History of USCGC *Dexter*," Washington, DC, undated.

U.S. Coast Guard Headquarters Public Information Division, "*Texaco Oklahoma* Marine Board of Investigation," Washington, DC, 30 March 1971.

U.S. Coast Guard Marine Board of Investigation, "SS *Texaco Oklahoma*, loss of in Atlantic Ocean off Cape Hatteras, North Carolina on 27 March 1971," Washington, DC, 26 October 1971.

White, Robert M., Administrator, National Oceanic and Atmospheric Administration, "Closing Down Ocean Station Hotel," letter to Adm. Owen Siler, Commandant, U.S. Coast Guard, Washington, DC, 18 March 1977.

CHAPTER 11

Books and Periodicals

Waters, John M., Jr., Captain, USCG (Ret.), *Rescue at Sea*, 2nd ed., Naval Institute Press, Annapolis, MD, 1989.

Government Documents

National Oceanic and Atmospheric Administration, National Data Buoy Center, Buoy History. Available at: http://www.ndbc.noaa.gov/ndbc.shtml. Accessed 1 June 2009.

National Oceanic and Atmospheric Administration, National Data Buoy Center, Chart of Buoy Locations. Available at: http://www.ndbc.noaa.gov/wstat.shtml. Accessed 1 June 2009.

National Oceanic and Atmospheric Administration, History Timeline. Available at: http://www.history.noaa.gov/legacy/time1900_1.html. Accessed 1 June 2009.

U.S. Coast Guard Commander, Fifth District, "Press Release: Man Medevaced from Ship Far Offshore," Portsmouth, VA, 28 March 2007.

U.S. Coast Guard Cutter *Mendota*, Official Logbook, August 1955.

U.S. Coast Guard Cutter *Minnetonka*, Official Logbook, November 1957.

U.S. Coast Guard Headquarters, Office of Engineering, *Loran*, *Volume I*, *Early Electronic History and the Bridging of the North Atlantic and North Pacific*, Washington, DC, December 1944 [U.S. Coast Guard Headquarters Public Information Division, *The Coast Guard at War*, *Loran*, *IV*, *Volume I*, Washington, DC, 1 May 1946].

INDEX